W9-AQW-274

The Brain-Shaped Mind
What the Brain Can Tell Us about the Mind

Will brain scientists ever be able to read our minds? Why are some things harder to remember than others? Based on recent brain research and neural network modeling, *The Brain-Shaped Mind* addresses these and other questions, and provides a clear account of how the structure of the brain influences the workings of the mind. Neuroscientists are now learning about our minds by examining how the neurons in the brain are connected with one another and the surrounding environment. This book explores how neural networks enable us to recognize objects and learn new things, and what happens when things go wrong. The reader is taken on a fascinating journey into what is arguably one of the most complicated and remarkable aspects of our lives.

Born in New York, NAOMI GOLDBLUM originally studied mathematics at Yeshiva University. She later moved to Israel where, at the Hebrew University, she extended her interests to the field of psychology. Her doctorate was entitled "A psycholinguistic study of metaphor." Naomi Goldblum is now a lecturer in psychology at Bar-Ilan University, where she specializes in cognitive psychology, in particular psycholinguistics and the processes involved in creative endeavors.

The Brain-Shaped Mind

What the Brain Can Tell Us about the Mind

NAOMI GOLDBLUM

Illustrated by
SHIFRA GLICK

CAMBRIDGE
UNIVERSITY PRESS

BOWLING GREEN STATE
UNIVERSITY LIBRARIES

PUBLISHED BY THE PRESS SYNDICATE OF THE UNIVERSITY OF CAMBRIDGE
The Pitt Building, Trumpington Street, Cambridge, United Kingdom

CAMBRIDGE UNIVERSITY PRESS
The Edinburgh Building, Cambridge CB2 2RU, UK
40 West 20th Street, New York, NY 10011–4211, USA
10 Stamford Road, Oakleigh, VIC 3166, Australia
Ruiz de Alarcón 13, 28014 Madrid, Spain
Dock House, The Waterfront, Cape Town 8001, South Africa

http://www.cambridge.org

© Naomi Goldblum 2001

This book is in copyright. Subject to statutory exception
and to the provisions of relevant collective licensing agreements,
no reproduction of any part may take place without
the written permission of Cambridge University Press.

First published 2001

Printed in the United Kingdom at the University Press, Cambridge

Typeface Lexicon (*The Enschedé Font Foundry*) 9/13 pt   *System* QuarkXPress™   [SE]

A catalogue record for this book is available from the British Library

ISBN 0 521 56104 3 hardback
ISBN 0 521 00094 7 paperback

Contents

Preface

I first heard of connectionism in 1982, when I began studying cognitive psychology. I had read Edward deBono's pioneering work, *The Mechanism of Mind*, twenty years earlier, and I had found it fascinating, but at that time the term "connectionism" had not yet been invented. When I learned about semantic networks, in which concepts were represented as points connected by links of various sorts, it seemed to me that concepts were much too rich to be described as mere points. Instead, I imagined them as long tangled threads meandering around in several dimensions, and I imagined the links between the concepts as the points where these threads met.

When I described this image to my cognitive psychology professor, Benny Shanon, he said, "That's the new theory everyone is talking about — it's called connectionism." He had just ordered the brand-new book on the topic, Hinton and Anderson's collection of papers, *Parallel Models of Associative Memory*, and was waiting for it to arrive. When the book came we spent a lot of time arguing over who should get to read it first. Each of us would take it home for a week or two and try to read a few pages, then give it to the other for the next week or two. On the one hand, the new ideas were fascinating, but on the other, they were very difficult to grasp.

Over the years since then I have read a great many papers on connectionism, but none of them was easy enough to recommend to a beginner. Even the few "introductory" textbooks that have been published on the topic require a great deal of prior knowledge, mainly of advanced mathematics and computer programming. After searching in vain for a clear, simple introduction to this difficult topic, I finally decided that I would have to write it myself. This book is the result of that decision.

Originally I thought of connectionism as a theory of the "mind–

brain," as if mind and brain were simply two sides of one coin. After long discussions with my philosophy professor, Avishai Margalit, I began to realize that mind and brain must be thought of as two separate entities, no matter how closely they may be entwined. The mind is not just an aspect of the brain, but a product of the interaction between the human organism and the environment. Still, the brain is a most important factor in shaping the mind, and so understanding the way this shaping takes place is of utmost significance in understanding human thought processes. The present book is thus devoted to this task.

I would like to thank my psycholinguistics professor I. M. Schlesinger and my research methods professors Maya Bar-Hillel, Ruma Falk and Yaakov Schul, as well as the two abovementioned professors, for teaching me how to think critically about both experimental results and theoretical pronouncements. My thanks also go to Benny Shanon and philosophy professor Avital Wohlman for comments on the manuscript of this book.

My daughter, Shifra Glick, not only read the manuscript but also drew the cartoons. I greatly appreciate her contribution. Various other family members and friends supported me throughout the writing of the book, and I thank them all.

Naomi Goldblum
Jerusalem

Figure permissions and acknowledgments

Figure 3.1: From Kolb *et al.*, *Fundamentals of Human Neuropsychology*. © 1996 W. H. Freeman & Co.
Figure 3.2: From Carlson, *Physiology of Behavior*. © 1994 by Allyn & Bacon. Reprinted by permission.
Figure 3.3: From Carlson, *Physiology of Behavior*. © 1991 by Allyn & Bacon. Reprinted by permission.
Figure 9.1: From Martin, *Neuroanatomy Text and Atlas*. © 1989 Pearson Education.

1

Introduction

How is the brain related to the mind? Do our minds work like computers? Can science's new knowledge about the brain tell us anything of importance about the way our minds work? How can a three-pound mass of tiny jelly-like blobs connected by vast numbers of microscopic filaments be the basis of all our thoughts, feelings, memories, hopes, intentions, knowledge? Will all the new knowledge scientists are gaining about our brains enable them to read our minds with electronic devices?

There is a new way of thinking about the mind and the brain which takes it for granted that the human mind is inseparable from the human body. Since the evidence indicates that the center of our mental activities is the brain, the advocates of this approach try to understand the functioning of the mind on the basis of what we know about the functioning of the brain. This new scientific paradigm has led to the construction of new theories and models of the mind which are variously known as connectionist theories, or neural network models, or theories of parallel distributed processing (PDP). Although some of the ideas on which these new theories are based have been around for over a century, the detailed working out of these models began only in the 1970s.

These new ideas are called connectionist theories because they claim that our mental processes and capacities – how we perceive what is out there in the world, how our knowledge about these things is organized, how we combine all this information to draw new conclusions, how we decide what to do next in order to get what we want – can be explained on the basis of what is known about the multiple interconnections between the neurons, or nerve cells, in the brain. They are called neural network models because they present detailed computer models of how interconnected units can work together to form networks analogous to those in the brain. And they are called theories of parallel distributed processing because they claim that a variety of mental operations are carried out at the same time, in parallel, and that these operations are distributed over large numbers of units rather than occurring within individual units separately. Each of these concepts will be explained in detail shortly, but first I would like to say a little more about the relation between the mind and the brain.

How are the mind and the brain related?

The connectionist view is based on the idea that there can be different levels of explanation for talking about the same thing. The concept of levels of explanation is well known in such disciplines as physics. For example, there is a difference between the level of our ordinary talk about tables and the atomic level of description. In our ordinary way of talking, a table is a solid object that entirely fills the space it is in and can have a smooth surface with no bumps on it. On the atomic level, in contrast, there is a great deal of empty space between the atoms that make up the table, and since these atoms are constantly jiggling around, there is no clear boundary between the top of the table and the air above it.

Similarly, there are both a mental and a physical level of explanation for talking about human mental functions. When we use the word "mind," we are on the mental level of explanation. It is on this level that we talk about seeing a sunset, remembering our trip to the Grand Canyon, knowing that a canary is a bird, and knowing how to tie our shoelaces. When we use the word "brain," we are on the physical level of explanation. On this level we can talk about individual nerve cells firing when they are activated by other nerve cells, the arrangement of nerve cells into columns in certain parts of the cortex, and the fibers that link one part of the cortex with another. But when we talk about red and blue color receptors being activated in the visual cortex, or the supplementary motor cortex sending electrical impulses to the primary motor cortex so that the muscles in our hands will contract in a particular way in order to cross one end of the shoelace over the other, we are combining two different levels of explanation of the same event – the mental and the physical, the mind and the brain. "Red" and "blue" are on the mental level of explanation, "receptors" on the physical level.

In the field of perception almost all scientists use the physical level of explanation in trying to understand the mental one. At least part of the explanation of the way we see requires an understanding of how the brain processes the signals coming in from our eyes. In contrast, many scientists working in the field of the higher mental processes – cognitive scientists, who study such topics as memory, language processes and the organization of concepts – claim that the mental level of explanation can proceed independently of the physical level. Most of them agree that the mind is inseparable from the brain, but they do not believe that it is necessary to ground the mental level of explanation in the physical one. Some cognitive scientists also claim that mental explanations of cognitive processes should be the same whether the processes are taking place in a human or a computer, for example; these researchers are often called students of artificial intelligence, or AI for short.

In contrast, connectionists believe that it is helpful to make use of our knowledge about the physical workings of the brain in our explanations of cognitive processes. The connectionist level of explanation may be thought of as a third level, intermediate between the mental and the physical ones. It is not identical to the physical level because it does not talk about individual nerve cells firing and is not concerned with the physical layout of the various parts of the brain. It is not identical to the mental level because it does not talk about our concepts, say, as abstract things that could be found in any entity that is able to process information. Rather, connectionists propose a description of mental processes that takes account of the physical structure of the brain and its interactions with the environment. As we shall see throughout the book, they try to explain how our concepts are formed, how they are related, and how they are used by looking at how the neurons in the brain are connected with one another and with the surrounding environment. Connectionists' explanations of mental processes are thus based on what we know about brain processes in much the same way as physicists' explanations of the different properties of wood and glass tables are based on the atomic composition of these materials.

How are connectionist explanations different from other explanations of mental processes?

The key to the difference between connectionist explanations and those given by other cognitive scientists, some of whom like to call themselves "classical," lies in the words "parallel" and "distributed" of the expression "parallel distributed processing." Classical explanations of mental processes describe them as taking place serially, one after the other. A good example is what is often called "memory search." When you actively and consciously search for a given memory, say, where you left your credit card when you notice that it is missing, you do a serial search: "I started out at the bookstore, then I went to the record shop, then to the video rental. Since I was able to use my card in all these places, I must have left it at the last one – the video rental." This is the way conscious memory searches take place, because we can focus on only one thing at a time, so we have to do our thinking serially.

But most of the time we seem to remember things without any conscious search of our memory. You see an unusual fruit, try to think of its name, and eventually "guava" just pops into your mind without any awareness of a search. Classical descriptions of this sort of remembering are modeled on conscious memory searches, due to an implicit assumption that the two types of remembering are similar. According to the classical theory, what happens when you see this unusual fruit is that you unconsciously go through a list of all the fruits you know and check each one to see if it matches the one you are looking at. This would have to occur very fast, of course, since "guava" pops into your mind fairly quickly, but in theory there could be a series of very fast processes. In contrast, connectionists claim that the names of the fruits you know are all activated to some degree when you see the guava. The more common fruits are activated more quickly, so it takes a while for the sight of the guava in front of you to make the word "guava" come to the forefront, but eventually it does. This is called "parallel processing" because the names of all the fruits are activated at the same time, in parallel, and no serial search is needed. This is explained in detail in the central chapter of the book, Chapter 5, "What are connectionist networks?"

The other important difference is embodied in the word "distributed." In classical explanations every concept, including "guava," is stored in one particular place in the mind. "Guava" is associated with "tropical", "fruit," and its various other features, but these are separate concepts, as I explain in Chapter 4, "Theories and models of how the mind functions." Connectionist explanations, in contrast, claim that every concept is made up of many parts, so that "tropical" and "fruit" are actually part of the concept "guava" rather than just being associated with it. This idea is explained in detail in Chapter 5.

These different modes of explanation have had different degrees of success in explaining different sorts of mental functions. The degree of success is often judged by how well the mental process can be modeled by a computer program, as I explain in Chapter 4. Classical cognitive scientists have had great success in making computer models of the sort of things human beings generally find difficult, such as playing chess, solving mathematical problems, diagnosing rare diseases, or finding flaws in complicated machinery. These are processes that we often know how to describe, things we do with a good deal of conscious awareness of how we do them. Thus experts in these areas could tell classical computer scientists how they did these things, and the computer people could program their computers to do them just as well – or even better.

But when the computer scientists tried to get their digital computers to do things humans do easily – such as making sense of simple stories with obvious details left out, or looking at a two-dimensional picture and seeing individual three-dimensional objects in it, or carrying on a sensible conversation – they had a very difficult time. Part of the reason is that we don't really know how we do these things, since most of the work is done outside conscious awareness. Another part of the reason is that neither the brain nor the mind works like a digital computer.

The brain isn't programmed by anyone; it grows and develops the abilities it has. The brain does come equipped with its basic structure, but this structure is constantly being changed by our experience in the world. This is not like changing a program on a computer, where the "hardware" remains the same but the "software" changes. In the brain there is no distinction between "hardware" and "software." Every change is a change of hardware. Every time we learn something new, every time we see a new scene, hear a new sentence or tune, touch a new fabric, taste a new dessert, smell a new flower, the connections between the neurons in our brain undergo some changes, and this constitutes a change in what we know – a change in our mind.

Connectionists also use computer models, but they try to make them work in a parallel distributed way. This involves trying to make computers be more like the brain instead of assuming that the brain works like a computer. Chapter 8 describes some models of this sort.

Is the mind in any way like the Internet?

Instead of comparing the mind to a computer, it might be more useful to compare it to the Internet. I am not saying that the mind is actually very similar to the Internet, but only that there are some interesting aspects of the mind that can be appreciated more easily by comparing them with properties of the Internet.

One of the most interesting properties of the Internet is the fact that it is not hierarchical – it is not controlled at the top by anyone who tells everyone else what to do. Instead, every part of the Internet can communicate with every other part, sometimes directly and sometimes through one or a few intermediate stations.

Connectionist theories claim that the brain, and therefore the mind, works this way too. The various parts of the mind – its individual networks – are all interconnected in a vast web, each part of which can communicate with any other part. In most of the book I will be describing the way the individual networks operate and how certain ones communicate with other ones, but it is important to remember that the flow of information can go in all directions. There is no "master" operator – no "DOS" – that determines where the information should go or what should be done with it when it gets there.

The mind does indeed have what is often called an "executive function," which is responsible for such activities as planning ahead. But although this function is generally in charge, it too can be overridden. My "executive function" may have determined that I will read a chapter of my physics textbook this evening, but if my external senses tell the sensory parts of my brain that the waterpipes in my kitchen have just burst, or if my internal sensors tell the proprioceptive parts of my brain – the parts that monitor internal states – that I am intensely thirsty, or if the person I love tells me that we need some time together, I may temporarily abandon my carefully thought-out study plans. And in case you are thinking that there is some "super-executive" function which decides that the flooding emergency is more important than studying physics, just think of the situation where you stop studying in order to watch a silly comedy on television. Different parts of your mind are in conflict here, and which one wins is not decided by some "super-judge"; the winner is simply the one that manages to gain the upper hand at that particular moment.

Thus the "executive function" too is just one part of the mind among many, all of which talk to each other incessantly and jointly direct our actions. This is the most important way in which the mind is like the Internet.

What about consciousness?

The idea that there is no hierarchy in the mind is very similar to a notion expounded by Daniel Dennett in his book *Consciousness Explained*. In this book Dennett demonstrates that there is no one part of the brain in which consciousness is "located," because all the various parts work together to produce our conscious experiences. The connectionist theories I describe in the present book fit in well with Dennett's ideas. However, they center on specific areas of the brain, how they are organized internally, and how they interact with other specific areas. They have not yet reached the point where they can come to grips with the interplay of processes occurring in many parts of the brain that probably underlies consciousness. Therefore I will not be discussing the topic of consciousness in this book.

It is very likely, nevertheless, that someday connectionism will have something interesting to say about consciousness, since I do not think that it is some mysterious entity that cannot in principle be explained by science. I am convinced that consciousness is embodied primarily in brain function, just like all other aspects of the mind. In fact, there are a number of possible scenarios that I can envisage if connectionists ever do try to explain consciousness. They may end up agreeing with Dennett that the concept of consciousness is not a particularly useful one for understanding the mind. They may come to the conclusion that it is such a complicated function of all the networks acting together that it is too difficult to explain. They may find out that it is grounded in a different part of the brain from the ones that hold the knowledge networks to be described in the following chapters, and so operates in a different way. But whatever may turn out to be the case for consciousness, I find that it is possible to understand a great many things about how our minds function without considering it. I shall therefore put aside the problem of consciousness in the present book.

What about the emotions?

The mind is sometimes considered to include the emotions as well as the intellect. In this book I will be discussing only the intellect, not the emotions. I will use the word "mind" rather than "intellect" because "mind" is the word that people generally use when they are speaking about such human functions as classifying things into categories, speaking, understanding speech, solving problems and the like.

It is possible that connectionist theory may turn out to apply to the emotions as well as the intellect, but it is also possible that it may not apply to the emotions. The reason for this is rooted in the structure of the brain. The part of the brain that is organized into neural networks is the cortex – the deeply folded outer part – which is primarily responsible for perception, thinking and the planning of action. Therefore it makes sense to use the properties of these neural networks to try to understand the mental functions that involve mainly the cortex.

The emotions, in contrast, involve primarily the inner structures of the brain, which are not organized in quite the same way. Thus rather different models may be needed to describe how the emotions work. Perhaps network theory will someday be able to shed light on the human emotions as well as the human intellect, but it has not yet done so. The discussions in this book will therefore be limited to the intellectual functions of the mind.

What questions does this book try to answer?

Some of the questions about our mental processes that connectionist theories offer answers to, which I discuss in this book, are listed below.

Will scientists ever be able to read our minds? If we understand the brain completely, will we be able to know what other people are thinking by looking inside their brain? In what ways are the minds of different people similar, and in what ways are they different? These questions are discussed in Chapter 2.

How is the brain put together so as to serve as the basis for our mental functions? What are the physical connections that allow us to form mental associations between the different things we know about the world? What are the physical bases of the changes in our minds that constitute learning? These are some of the questions discussed in Chapter 3.

Chapter 4 discusses the following questions: What is the difference between a theory and a model? What models were used to explain the organization of our knowledge before connectionist theories were developed? How can computers help us understand the way our minds work even though the mind doesn't work like a computer?

The essence of connectionist theory is discussed in Chapter 5, with emphasis on the following questions: How do we put things into categories? How do we know, for example, that the animal we are looking at is a dog and not a cat? How are all the things we know about various animals connected with each other, so that we can instantly produce a long list of properties of any animal, or provide the name of the animal from a description of it?

Another central problem is how these connections form in the first place. How do children learn the difference between a cat and a dog? Why do some children call all four-legged animals "dog" or "horse" at first, although they never call a bird "horse"? How do they later learn to distinguish between different types of dogs, and even between individual dogs? These are some of the questions addressed in Chapter 6.

Of course, not all our mental associations are between objects and their names or physical properties. We also know that various things are related, such as mothers and fathers, or salt and pepper, or even properties of things, such as high and low, or sweet and sour. But how do we know in what way these things are related? How do we know that a wolfhound is a kind of dog rather than a kind of wolf? Why do we say "dog" and not "tiger" when we are asked for a word associated with "cat," even though cats and tigers are more closely related than cats and dogs? These questions are discussed in Chapter 7.

What sort of experimental evidence is there that the models discussed in this book actually describe what goes on in the human mind? Chapter 8 describes some of this evidence, including computer models of how children learn to talk and how they learn the past tenses of verbs.

Chapter 9 discusses the difference between things we remember for a long time and things we remember for only a little while. For example, how do we know that today we had eggs for breakfast, even though we usually have cereal, and why is it that next year we will still remember that we had cereal for breakfast most of the time this year, but we will not remember that we had eggs today? How do we remember before we go to the bank that we have to go to the bank this morning, yet instantly afterwards remember that we already performed this errand?

All the questions so far have involved the normal functioning of the mind in the intact brain. What happens if this functioning is disrupted by damage to the brain, such as that caused by strokes, head injuries or degenerative diseases like Alzheimer's? Chapter 10 discusses the different sorts of dysfunction associated with each of these causes and the prospects for regaining normal mental functioning in certain cases even if the brain damage itself cannot be repaired.

Is there any way we can use all this knowledge to improve our mental functioning in our daily lives? Can it help us study better, or teach others

more effectively, or solve our personal problems? These and other practical implications of connectionist theory are discussed in Chapter 11.

Neural network models are still fairly speculative at this point. Although the experimental evidence supports the theory that the brain and the mind work in this parallel distributed manner, the evidence that the mind works this way is rather less certain than the evidence with regard to the brain. As a result, there has been much criticism of connectionist theory, particularly by people who advocate other theories about mental functioning that are less closely tied to the workings of the brain. A sampling of these criticisms is presented in Chapter 12, together with replies that have been offered by researchers in connectionist theory.

Despite the fact that connectionism has its critics, it does reflect one of the most prevalent ways of thinking about the mind at the present time. In my discussion of what the models imply about how the mind works, I will therefore assume that they are true, and present their exciting implications by saying simply that this is what our minds do, rather than continually repeating that this is what the models say that our minds do.

In essence, although I describe a fairly simple neural network model in some detail, my discussion of this model is not meant to be an end in itself. I see it, rather, as a way of showing how our developing understanding of how the brain works can help us understand some aspects of how the mind works as well. Brain processes are only one of the forces that shape the mind – it is also shaped by input from the environment, both physical and social. Describing how these two types of forces interact requires another book entirely.

I have tried to keep the descriptions in this book clear and the explanations simple, without losing sight of the essential points of this complex subject. I hope you will enjoy this journey into what is not only the most complex but, for me at least, also the most fascinating object in the universe.

2
___

What the brain cannot tell us about the mind

Why should we base our theories about the mind on our knowledge of how the brain functions? Why shouldn't we study the mind as philosophers and scientists did for centuries, by introspecting to find out how we ourselves think and then asking other people questions to find out how they think? Or if we want to be more scientific, and we can figure out how to make a computer do what human beings do, why not assume that the way the computer does it is the way humans do it too?

One reason is that there is a very great difference between our ability to know what we are thinking and our ability to understand how this thinking takes place. We are aware mainly of the contents of our thoughts, because this is the knowledge we need in order to be able to function. I need to know that the object I am looking at is an apple in order to know that I should pick it and eat it rather than the leaf next to it. I do not need to know how I recognize it as an apple in order to eat it and be nourished. Our consciousness has therefore evolved to be aware of *what* we are thinking, but not of *how* we go from one thought to another, or how we go from a perception to a thought, or from a thought to an action.

But our minds at some point, perhaps at the time of the Greek philosophers, began to consider the way they themselves work. As human beings, we cannot be satisfied with just knowing facts about the world "outside" ourselves. We want to understand how we think, how we come to know what we know. The world "inside" our heads seems very different from the world "outside," and we long to find out how it operates. But since we are not built to be able to get at this knowledge directly, we have sought all sorts of indirect ways of finding it out.

Dualists and monists

Many philosophers – most notably Descartes – have been so impressed by the differences between what we sense when we observe the "outside" world and what we sense when we turn "inward" to discover what is in our minds that they are convinced that the "outer" and the "inner" constitute two totally separate "worlds." This is the dualist approach – the belief that mind and body are two totally different aspects of existence. Dualists believe that since the brain is a physical entity it has nothing to teach us about the nonphysical mind. And because most of them do not distinguish between the contents of our thoughts and the processes of our thinking, they believe nothing we can learn about the brain will ever be able to explain how our minds work.

Monists, on the other hand, believe that the mind and the brain are one and the same, looked at from two different aspects, so that advances in the study of the brain will eventually teach us all we need to know about the workings of the mind. Many of them also do not make the distinction between the contents of our thoughts and the processes of our thinking, and they believe that there is nothing about the mind that would remain unknown if we knew all there was to know about the brain.

To me, however, this distinction seems crucial. I believe that we can learn a great deal about how our minds work, about the *processes* of our thinking, by studying the workings of the brain. The *contents* of our minds, on the other hand, will always need to be studied separately. In other words, while the mental level of explanation of our cognitive processes can usefully be anchored in the physical level of explanation, there is no point in trying to explain our particular pieces of knowledge or belief in this way. I will now try to make this distinction clear.

The difference between process and content

What we can learn about the mind from studying the brain is how the mind operates, in the sense of how it processes information. Scientists are now learning more and more about the way neurons send information to one another, the way they are organized into networks, the way these networks incorporate new information and the way they make use of the information they already have. This new knowledge about the brain can be used for building models of how we recognize people and things, how we classify things into categories, how we learn to do things we couldn't do before, how we learn to speak and later to read, and how we act on the basis of what we believe and what we want. What we cannot learn from studying the brain is what information there is in the mind – what you or anyone else is thinking at any particular moment, or what knowledge you or anyone else possesses in general.

The rest of this book is devoted to those aspects of the mind that I believe we can understand better by using our knowledge of how the brain works. In this chapter I explain why I nevertheless believe that no one will ever be able to find out what you are thinking or what you know by finding out what state your brain is in, no matter how much we learn about the brain and how well we are able to monitor what your brain is doing.

Token–token identity

The particular belief that I hold about the relation between the mind and the brain is a sort of monism, since I believe that the mind is inseparable from the brain. The sort of monism I accept is based on the difference between types and tokens. For example, a cat is a type of animal, while an individual cat – say, your pet Lucky – is a token of this type. Some facts are true about cats in general, while others are true only about Lucky. The facts about cats in general can be considered scientific laws about cats – for example, "All cats have fur." Facts about a particular cat – say, that Lucky wears a purple ribbon with her name printed on it in yellow letters – do not have the status of laws.

What does this distinction between types and tokens have to do with the mind and the brain? It marks the difference between two ways of thinking about the relation between mind and brain. One of these, called the "type identity thesis," claims that this relation can be described in terms of general laws. In other words, for every type of mental event – all stabbing pains in the left big toe, all thoughts that the grass is yellow, all intentions to go to Alaska, all wishes to be a millionaire – there is a particular type of brain state or brain process that corresponds to it in every person who is experiencing this type of mental event. This means that the mind can be "reduced" to the brain – that is, once the brain has been described in physical terms we will know everything there is to know about the mind.

A different way of thinking about the relation between mind and brain is called the "token identity thesis." Although the advocates of this thesis also believe that the brain is the basis substrate of the mind, they do not agree that each type of mental event corresponds to a particular type of brain state. What they claim, instead, is that there is a correspondence between tokens of mental states and tokens of brain states – that every time we think or feel or sense or want something there is some process occurring in our brain. In other words, when I see a red square and you see the same red square, neurons are activated in the visual areas of my brain and neurons are also activated in the visual areas of your brain. However, the neurons firing in my brain when I see a red square are probably not the same as the ones that are firing in your brain, in the sense of "same" that we use when we say that if I am typing the word "red" on my PC and you are typing the word "red" on your PC with an identical keyboard we are both depressing the same keys in the same order. Similarly, if I am thinking that Arafat needs a shave and you are thinking that Arafat needs a shave, it is most unlikely that the neurons activated in my brain are the same as the ones that are activated in yours. Thus the correspondence between mind and brain involves only tokens, not types.

Why no brain-scanning machine will ever be able to read minds

This is why I believe that there is no need to worry that someday scientists will be able to find out what we are thinking by hooking us up to machines that can tell them which neurons are active in our brains. There are, to be sure, machines that perform PET scans, which can tell what sort of processes are going on in our brains. These machines produce pictures that show in vivid color which parts of our brains are most active when we are performing a mental calculation or watching a movie or listening intently to our favorite music. Then why am I so certain that further scientific progress won't enable us to pick out the exact neurons that are active when we are thinking that "2+2=4" or watching *Casablanca* or listening to the Beatles' rendition of "Yesterday"?

To explain why I am so sure that this will never happen, I will detail my reasons for believing that the neurons that are active in my brain when I am thinking that "2+2=4" are not the same as the ones that are active in yours when you are thinking the same thing. If this is the case, then it is enough to ensure that no one will be able to "read your mind" by "reading your brain." Let's assume for the sake of the argument that scientists will someday be able to pick out the exact neurons that are firing

"Gotcha!"

in my brain when I am thinking that "2+2=4" or when I am seeing a red square or when I am intending to get some chocolate ice cream out of the freezer. But if the neurons that are firing in my brain when I am thinking that "2+2=4" are not the same as the ones that are active in yours when you are thinking the same thing, then it won't do the scientists any good to know which ones are firing in my brain, because even if they could see the "very same" neurons firing in your brain – that is, if they could find neurons active in exactly the same physical location in your brain as in mine – they would not be able to deduce that you are also thinking that "2+2=4." Although they might be able to deduce that you are thinking about some numerical fact because the "numerical" area of your brain is lit up, they wouldn't know which particular numerical fact you are think-ing about. Similarly, if they discovered exactly which neurons are firing in my brain when I am thinking that Arafat needs a shave, and they saw the "same" neurons firing in your brain, they might be able to say that you too are thinking about something related to the visual appearance of a human face, but they could not deduce that you are thinking that Arafat needs a shave.

For a more detailed example, consider the brain areas which work together to enable us to speak, since there has been a great deal of progress in our understanding of these areas. Not only do we know which areas of the brain are responsible for speech, we can even distinguish the areas that are activated when you just have a general idea of what you are going to say from the areas activated when you are actually forming the sentence out of individual words. For example, we know that if a certain area on the left side of your brain, known as Broca's area, is electrically active, then you are likely to be formulating a sentence that you are about to say, because this area is a key center in the process of formulating sentences for most normal people. For the small number of people whose language centers are on the right side of their brain, this is not the case, but the generalization is a useful one anyway because it is true for most people.

Yet it will never be possible to use our knowledge of the brain to predict the exact sentence you are formulating when your Broca's area is lit up. We might be able to say something about its general content – if, for example, the area responsible for mathematical calculations was lit up a fraction of a second earlier, then it is very likely that you are about to express the result of some such calculation. But we will never be able to predict, say, that if neurons G3, W1243 and X756 are active, then you are about to say "2+2=4," while if H856, Q2064, V902 and three hundred other particular neurons are active, you are planning to say "If it takes Faucet A 20 minutes to fill up the bathtub, while Faucet B can do the job in 30 minutes, then if both faucets are turned on the bathtub will be filled up in 12 minutes."

Therefore, even if "brain-readers" could map all my present thoughts onto specific brain states by recording which neurons are firing each time I am thinking some specific thought, this would enable them to know only what I myself am thinking when the same configurations of neurons fired again, and then only in case I had not learned something new about this particular topic in the meantime. They would not be able to deduce the specific details about what anyone else was thinking if the same configuration of neurons fired in some other person's brain, because that configuration might be used to encode some other thought.

But science is built on generalizations. Being able to read one specific person's mind would not be worth the effort, as it would be easier just to ask her what she is thinking. The effort of mapping neuronal configurations onto thoughts would be worthwhile only if this knowledge could be used to read other people's minds as well, but what I am claiming here is that this is impossible.

Why is this so? If neuroscience has advanced to the point that we can tell whether you are making a mathematical calculation or imagining a colorful scene, how can I be so sure that further advances in this area will not allow us to figure out which particular mathematical calculation you are making, or what scene you are imagining?

The detailed structure of everyone's brain is different

The answer is not that the technology couldn't be invented, since there don't seem to be any limits to the sorts of technology that can be invented. It is rather that there wouldn't be any point in it, because each person's knowledge is organized differently within her/his brain, so that no generalizations can be made on the level of specific pieces of information. Even if we could use some sort of advanced technology to discover precisely which neurons in Dick's brain are active when he is getting ready to say "See Spot run," it wouldn't tell us anything about which neurons are active in Jane's brain when she is preparing to utter the very same sentence.

There are at least two reasons for this – one connected with heredity, and the other with environment. As in every other area where heredity and environment are involved, the two actually interact in very complex ways, but in order to understand their interaction we must first separate out the two aspects.

The heredity-based reason is that the fine structure of the connections between the neurons in different people's brains is different, just as everyone has different fingerprints from everyone else. Normal people all have their middle finger longer than their other fingers, for example, but the exact pattern of whorls on the fingertip is different for each person. We can, of course, take everyone's fingerprints and thus know the exact pattern of whorls on every person's fingers, but that will not enable us to predict the exact pattern of the next person's fingerprints. In the same way, some new advanced technology might someday enable us to find out which neurons are active in Dick's brain when he is about to say "See Spot run," but this will not tell us anything about which neurons are active in Jane's brain when she is planning to say the same thing, because the fine structure of the neurons in their brains is different. We may eventually find out that there is a particular *area* of the brain that is active when Dick and Jane, or anyone else for that matter, are thinking about animals, but the particular pattern of *neurons* that are active when they are about to say "See Spot run" is as individual as a fingerprint.

The environment-based reason for the difference between people in the exact pattern of neurons involved in planning to say "See Spot run" is that each person learns the individual words in a different way. The way we understand each word we know is based on the different sentences in which we've heard it used, and the different things that were going on at the time we heard these sentences. But no two people have heard exactly the same sentences, and even in the case of those sentences that were the same, not exactly the same thing was going on when they heard them. Thus the neurons that encode the word "run" are connected differently for Dick, who first encountered the word when his older brother yelled "Look at those dogs run" while taking him for a walk in the park, than they are for Jane, who first heard the word when her mother said "Don't run so fast." Therefore the particular pattern of neurons that encodes this word in Dick's brain cannot be exactly the same as the pattern that encodes it in Jane's.

But everyone's brain functions similarly

If no advances in our knowledge of the brain can tell us *what* exactly is going on in anyone's mind, how can this knowledge nonetheless help us understand *how* the mind works? I will try to explain this with an analogy and a counter-analogy. The philosopher Jerry Fodor, who believes that knowledge of the brain is unnecessary for understanding the mind, makes use of an analogy about money. Every coin or bill that constitutes money is a concrete, physical piece of metal or paper. However, the value of these coins and bills depends on the mathematical relationships between them, not on their physical form – a hundred-dollar bill is worth twenty five-dollar bills, even though the physical difference between a hundred-dollar bill and a five-dollar bill is very small. Therefore most of our economic laws are based on these mathematical relationships rather than on physical facts about coins and bills. In the same way, Fodor claims, facts about the physical structure of the brain are generally irrelevant for understanding the functioning of the mind.

There are, however, some generalizations about money that are based on its physical form. For example, paper bills are generally worth more than coins. Similarly, there are many interesting facts about our mental processes that can be explained at least in part on the basis of the physical processes taking place in the brain, and these facts are the subject matter of the present book.

A simple example may demonstrate how an aspect of the physical world can help us understand an aspect of the mental, even though it does not explain the mental phenomenon entirely. We know that light is a form of electromagnetic radiation, with different colors corresponding to different frequencies of this radiation. The radiation is the physical aspect of light, while the mental aspect involves the different colors we see. Nothing in our knowledge of electromagnetic radiation can explain the "blueness" of blue or the "redness" of red. These are essentially mental properties and will remain so no matter how much we know about the physical structure of light. But knowing the structure of the radiation can help us understand the structure of our color vision. The fact that blue and green look more similar to each other than red and green is explained quite simply by the fact that the frequencies corresponding to blue and green are closer to each other than the frequencies corresponding to red and green.

It is in this way that our knowledge about the brain can be used to help us understand how the mind works. Finding out how neurons are arranged, how they are connected to one another, how one neuron sends a message to another neuron, can help us build models of how information is processed in the sort of brain that human beings actually possess, as opposed to, say, how information is processed by digital computers, or how it might be processed by some hypothetical rational beings from Alpha Centauri. Let us take a look, then, at the insides of our brains.

3

How neurons form networks

What is it about the human brain that allows it to be the basis for all the complexities of the human mind, including not only language, reasoning and memory but emotion and intuition as well? How can this apparently small organ, weighing an average of three pounds, effortlessly accomplish many tasks, such as face and voice recognition, that are quite difficult for extremely sophisticated computers?

The basis for all these remarkable accomplishments is the great complexity of the connections between the elementary units that make up the brain – the neurons, or nerve cells. First of all, there are billions of neurons in the brain. Second, each individual neuron makes contact with about ten thousand other neurons, so that the actual number of connections between the neurons in the brain is astronomical.

Moreover, there are at least two different levels of connections involved. The brain can be subdivided into a number of modules – large-scale units consisting of some tens or hundreds of thousands of neurons. We do not know exactly how many such modules there are, but a fair estimate would be on the order of hundreds. Within each of these modules all the neurons are connected with all the other neurons, either directly or at one or two removes. It is because of the intricate connections among the individual neurons in each module that we call the modules neural networks.

How do these modules work?

Each module or network of this sort is responsible for one aspect or stage of some particular mental process, such as recognizing familiar faces or finding the right words for the sentences you are planning to say. In fact, since the system as a whole is so intricately interconnected, the same module often plays a part in more than one process. For example, there is evidence that some of the modules involved in seeing objects in the world are also involved in forming mental images of these objects. Likewise, the modules involved in understanding sentences also play a part in the process of producing sentences, so that we can monitor what we are saying to make sure that it isn't total gibberish.

In order for each individual module to play its part in each of the processes it is involved in, it must be connected to a number of other modules. For example, the face-recognition module must receive information from the shape- and color-perception modules, and must in turn convey its information to the module that contains the names of the people whose faces we recognize as familiar. The word-finding module must receive information from some planning module, and must interact with the module that arranges the words into grammatical sentences.

The interaction of the word-finding module with the grammar module illustrates the complexity of the connections between the various modules. They are not connected in only one direction – planning sentences to finding words to stringing the words together to uttering the sentences. There is a great deal of feedback along the way – some of the words for a sentence may be chosen before the grammatical structure is selected, but other words, especially function words like "in" or "but," may well be chosen only afterwards. Moreover, many of the feedback loops involve several modules, not merely two of them.

Thus the complexity of our thinking is embodied in a doubly complex tangle of neurons in our brains – the local complexity of the thousands of units within each network and the overall complexity of the feedback loops between the networks. In order to understand the complexity of our mental processes, therefore, we must first have a basic understanding of the workings of the neurons and modules in the brain. Neurobiologists have discovered a great many facts about the workings of the individual neurons in the brain, and how each one sends its messages to the next ones. I will present here only those facts that are most relevant to our understanding of the process of communication within neural networks. In contrast, much less is known about how cells work together to form networks – how they work in concert as large groups. Here I will present some of the latest and most exciting discoveries and theories in this area – the ones that form the basis for the belief that every individual piece of information is stored in a whole network rather than in an individual nerve cell, while every mental process requires the interconnected activity of a number of networks.

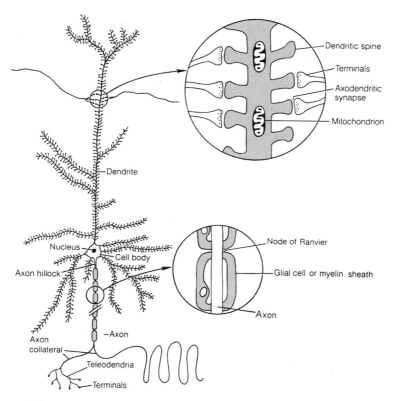

Figure 3.1. Schematic drawing of a typical neuron, or nerve cell. The inset at the upper right shows the synaptic gaps between axonal branch terminals of several sending neurons and dendritic tips of the receiving neuron.

How do neurons interact?

The first question is how the individual neurons operate. A schematic drawing of a single neuron is shown in Figure 3.1. The round body of the neuron is known as the cell body or soma. It serves all the usual functions of cell bodies – manufacturing energy for the cell's work from appropriate chemicals that it takes in from the blood stream, excreting wastes, manufacturing the specific chemicals that this particular cell is meant to produce. In addition, the membrane or outer covering of the neuron's soma collects electric charge, somewhat in the way a rechargeable battery does.

The dendrites, as you can see in the figure, are thin filaments forming a bush-like structure. Each of these dendrites can conduct a weak electric charge along its membrane to the body of the cell. If enough of these weak charges arrive at the cell body within a small interval so that its membrane becomes charged to a critical point, it will discharge its electricity along the long filament leading out of the cell, which is known as the axon. At the end of the axon are branching filaments which form bushes similar to those on the dendrite side. The important difference is that the axonal branches conduct electricity outwards from the cell body, while the dendrites conduct it inwards to the cell body.

When scientists first realized that electricity is conducted within nerve cells, they believed that these cells also send messages to one another by means of electric currents. This has turned out to be true of only a very few types of neurons. In general, it has been found that there is a gap, known as a synapse or synaptic junction, between one cell and the next. Although an axonal branch leading out of Cell A may come very very close to a dendrite leading into Cell B (see the enlarged inset at the upper right of Figure 3.1), the two do not touch and no electricity is conducted from one to the other. Since this is the case, how does the message get across the gap?

The way the message gets across is by means of a chemical known as a transmitter. There are several types of chemical that function as transmitters, but the differences between them are not important for our purposes. These transmitters are stored in little sacs called vesicles near the tips of the axonal branches. When the electric current reaches the tips of the branches it causes the vesicles to fuse with the membrane covering the branch tips, which then opens, so that the molecules of the transmitter spill out into the fluid between the cells (see Figure 3.2). These molecules then spread out in the fluid, and since the gap between the branch tips of Cell A and the dendrite tips of Cell B is very narrow, many of the molecules hit the membrane of the Cell B dendrites.

The dendrite membrane is studded with big protein molecules called receptors, which are shaped so that the molecules of the transmitter can fit into them the way a key fits into a lock. As each transmitter molecule is caught up into a receptor molecule, it increases (or, in some cases, decreases) the electric charge on the dendrite membrane by a small amount. It does this by causing pores to open in the cell membrane, thus allowing charged particles (ions) to flood into the cell. Increasing the charge on a neuron's membrane is called activation, while decreasing the charge is called inhibition.

The charge on the membrane is then conducted down the dendrite to the cell body, where it combines with the charges from other dendrites. If

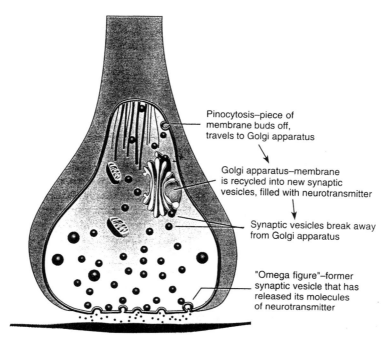

Pinocytosis–piece of
membrane buds off,
travels to Golgi apparatus

Golgi apparatus–membrane
is recycled into new synaptic
vesicles, filled with neurotransmitter

Synaptic vesicles break away
from Golgi apparatus

"Omega figure"–former
synaptic vesicle that has
released its molecules
of neurotransmitter

Figure 3.2. When the synaptic vesicles reach the membrane at the end of the axon tip, they fuse with the membrane and open up to release their molecules of neurotransmitter into the synaptic gap.

the total combined charge from all the dendrites is large enough, it will lead once again to a discharge of electricity down the axon, as described above. This cascade of activation continues over and over again to constitute the process of message transmission within the brain (see Figure 3.3).

Although the process always occurs in the same way, its individual stages contain aspects that can vary. For example, the amount of transmitter emitted by a cell can be greater or smaller. Moreover, the transmitter has to be taken back into the emitting neuron shortly after it is emitted, as otherwise the absorbing neuron would keep on being activated all the time, and not only when a message needs to be sent. This process of reuptake can be quicker or slower; the slower it is, the more strongly the second neuron is activated. In addition, the second neuron can have more or fewer receptors per unit area in its membrane at the point of the synapse. The more receptors, the more strongly the receiving neuron is activated. All these variables affect the strength of the synapses between the neurons.

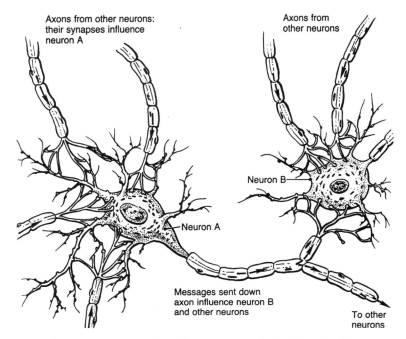

Axons from other neurons: their synapses influence neuron A

Axons from other neurons

Neuron B

Neuron A

Messages sent down axon influence neuron B and other neurons

To other neurons

Figure 3.3. Many axonal branches converge on the dendrites of each neuron, and the axonal branches of each neuron diverge to transmit their message to the dendrites of many other neurons.

How are neural networks formed?

How do the individual neurons work together to form networks? We saw that the charge produced by the transmitter fitting itself into the receptor of the dendrite membrane is quite weak. Moreover, it tends to dissipate within a very short time if it is not reinforced by many other charges arriving at the membrane during this interval. The dendrites of the receiving neuron must therefore get many more activating chemical messages than inhibiting ones within a short period of time if the cell body is to collect enough electric charge to send a current of electricity down the axon, thus passing the message on.

It is therefore important to know which axon branches synapse with which dendrites. If all the messages received by the dendrites of a particular neuron came from the axon branches of one other neuron, there would be no point in having so many dendrites and so many axon branches, since the message of Cell A would be picked up only by Cell B. The point of each neuron having so many dendrites and so many axon branches is that it can receive messages from very many neurons and send out messages in its turn to very many other neurons. Thus the axon branches spread out widely, forming synapses with dendrites of many different neurons.

But how does this help create a network? If each neuron sends messages to tens of thousands of other neurons, it would seem to yield a widely branching tree-like structure rather than a network. What creates the network is the fact that if Cell A transmits a message to Cell B, this does not prevent Cell B from transmitting a message to Cell A as well. Or Cell A might send a message to Cell B, which sends a message to Cell C, which then sends a message to Cell A. Such loops can exist with different numbers of neurons in the chain, and in a wide variety of combinations. Groups of tens or hundreds of thousands of neurons in which such message loops exist constitute the neural networks, the networks of the brain which serve as the basis for all our perception, our thinking, our memory, and the planning of our actions.

How are the modules connected?

The connections between neurons are not confined to those within individual networks. In order for the brain to process the information gained through perception and to plan actions to further our goals through the use of this information, the various modules must be connected with one another. In order for us to run away from a tiger, say, the neurons in the visual perception module must send a message about what it has seen to those in the module containing the images of animals, which in turn must send a message to the ones in the module that stores information about various animals. This module must then send its message about the tiger to the module that controls the actions of our muscles, which in turn sends the message to the muscles that tells them to run away. (This is, to be sure, only a very simplified sketch; there are actually many more modules involved, such as the ones in the limbic system which cause us to feel fear, and the ones in the cerebellum which co-ordinate the activities of our limbs.)

The connections between neurons in different modules are essentially the same as the ones between neurons in the same module, except that the axons are longer when messages are sent to another module – long enough to be able to reach the next module, in whatever part of the brain it may be.

The connections between the neurons keep changing

One of the most important features of the synapses between neurons is that they can change as a result of experience. This property of the synapses is called *plasticity*, as a (soft) plastic object can easily be shaped into different forms. Our next question is thus how the synapses change. This question has two parts: What causes the synapses to change, and how does this change take place?

Changes in the synapses occur according to a principle known as Hebb's rule: Every time a neuron fires after receiving an excitatory input from another neuron, the synapse linking the two neurons is strengthened. From the outset all the neurons in a network are connected with one another, whether directly or at one or two removes, as we have seen, and new synapses are rarely formed after the first period of neural development. Therefore almost all learning in the brain occurs through the strengthening or weakening of the existing links between the neurons. Each time one neuron provides part of the input needed for another neuron to fire, the synapse between the two is strengthened just a little bit. As a result, the second neuron is just that little bit more likely to fire the next time it receives an input from the first one, thus strengthening the synapse a little bit more. If this process is repeated over and over again, it results in very strong links between particular neurons.

As we have seen, messages are transmitted across a synapse in several stages, and the strengthening of the link can be accomplished by changes in any one or more of these stages. In the stage of transmitter release, more transmitter molecules can be released by the vesicles of the first neuron into the synaptic gap. The re-uptake of the transmitter molecules can also occur more slowly, so they have more time to act on the membrane of the receiving neuron. In addition, more receptors can be formed in the membrane of the receiving neuron, so that the impact of the transmitter will be greater. There are also other ways in which the synapses can be strengthened, but this sample should provide an idea of the variety of methods that can be used.

Weakening of synapses does not occur directly; there is no specific occurrence that can make a synapse weaker, in the way that it can be made stronger by the firing of the receiving neuron just after the firing of the transmitting neuron. Not very much is known about this process, but it has been speculated that connections that are never or rarely used gradually become weaker, by a reversal of some of the mechanisms that make the often-used connections stronger.

In subsequent chapters we will see how this whole system serves as the physical basis for the phenomena we call "mind." The synapses between the neurons store all our memories, all our plans for action, all our knowledge of the world, all our hopes and fears, and the changes in these synapses constitute all our learning, whether from formal schooling or from our life experiences.

4

Theories and models of how the mind functions

Now that we have some idea of how the neurons in the brain form networks and how these networks operate, we are in a position to take a good look at the question of how the mind works. Remember that we are assuming that the functions of the mind – how we learn new things, how we remember the things we have learned, how we combine the things we have learned to create new entities – are embodied to a great extent in the workings of the brain. The question we are now asking is how these functions of the mind are shaped by the paths taken by currents of electricity and wandering molecules within the brain.

The latest theories about how the functions of the mind are grounded in the operations of the brain are known as "connectionist" theories or models. The difference between the theories and the models that is important for our purposes here is that the theories are attempts to formulate general rules about how the mind performs its various functions, whereas the models are attempts to simulate these functions on computers. The theories are like all new scientific theories – they are systems of generalizations that are based on a new way of looking at accumulated observations which no longer seem to fit the old theories very well. The new generalizations are then examined to see how they could be tested, and experiments are carried out to see if the predictions made by the new theories are fulfilled. If the experiments give the hoped-for results, then we have more faith in the theory, and we try to elaborate it to cover more details, and expand it to new areas.

A model is a way of testing theories that is often used in place of experiments in conditions where actual experiments would be too costly, unethical or even impossible. For example, space flight simulators are used to see how weightlessness affects humans. We can't wait until humans are actually sent into space to see how weightlessness will affect them. If there are some severely harmful effects, we may not be able to get the people back in time to prevent these effects from getting worse. That's why we use simulators, such as huge tanks of water in which the astronauts are immersed to see what happens to them when they are effectively weightless for long periods of time. This way they can be taken out of the weightless environment immediately if anything goes wrong.

Computer models

Computer models of mental functioning serve a similar purpose. We can't open up people's heads just to see what's going on inside. Occasionally, it's true, we do have to open up some people's heads to remove a brain tumor. In cases like this, it is very important to have a way of finding out what functions are served by the area of brain tissue immediately surrounding the tumor, so that the surgeon will know which way to make the cut. For a purpose such as this it is clearly ethical to stimulate various neurons and see how the patient responds. (Brain operations of this sort are done while the patient is awake, using only local anesthesia for the skull, as the brain itself has no pain receptors.) However, there is a very limited amount of time available during the operation for asking the patient the questions essential for the proper conduct of the operation, and there is no way that the detailed observations necessary for testing theories about the mind could be made.

The recent development of PET scans does provide a way of finding out which modules of the brain are involved in the performance of various tasks, and this provides exciting corroboration of some connectionist theories. PET scans, however, can show us only the activity of large groups of neurons; they cannot tell us how the individual neurons activate or inhibit one another to form the network modules.

To be sure, cognitive psychologists, working from the perspective of the mind rather than the brain, have been performing experiments for several decades to see how some mental processes, such as recognizing and naming objects, remembering words and facts, reading words, and understanding sentences, are accomplished by the combined working of various modules in the mind. The experiments we can do to study such processes, however, are only of a few very simple sorts, such as asking different kinds of questions and seeing how long it takes people to answer them. This provides us with some information, but not enough.

Since none of these methods is capable of thoroughly testing the detailed theories about how the basic neural units work together to form modules and to provide communication among the modules, we also use simulations. One of the things we do is program computers to perform various functions in the way we believe the mind does. We connect up the units inside the computers similarly to the way our theory says neurons are connected in the brain, and we try to teach the system of computer units to perform some task the way we believe humans learn to do it. If the computer system ends up not only being able to do the same things humans can do but also doing them in much the same way, even making similar mistakes in the process, we then have a much stronger reason for believing our theories are true.

The computer models are based on our best knowledge of how the brain functions, but they go beyond the known data in their specific details. Although the past few decades have seen an enormous increase in our knowledge of brain processes, we still do not have the wealth of detail needed to set up functioning models on computers. Moreover, if connectionist modelers were to use data based on present knowledge of the brain, they would have to change their models every time a new piece of information was discovered. Therefore the connectionist models are based only loosely on what is known about the functioning of the brain. The models are broadly compatible with our current knowledge of the brain, but they do not commit themselves to specific details which may be proven false tomorrow.

An example may make this clearer. One of the most important areas where connectionist modelers do not commit themselves is the nature of the individual units that are connected up to form the networks. For simplicity I discuss the models as if the network units are analogous to single individual neurons, but this is not essential. It is entirely possible that we may one day find out that the units are actually small groups of neurons working together, or, conversely, subparts of single neurons. Since connectionists have not committed themselves to the equation of network units with individual neurons, such a discovery would not destroy connectionist theory. It is easier to speak as if these units are neurons, and I will therefore do so throughout this book, but it is important to keep in mind that the functional units may turn out to be either greater or smaller than a single neuron.

Reading words

Since most models of connectionist theory require some basic under-standing of the theory, I describe several such models in Chapter 8, after presenting the fundamentals of the theory in Chapters 5, 6 and 7. Here I will only describe one type of situation in which such a model might be used, to provide a concrete illustration of their purpose.

Consider two popular theories of how people read words. One is the "phonics" theory, which says that people first try to sound out the letters in a word and read the word on the basis of the individual sounds; only if this does not work do they search their memory to see if this particular word has an exceptional pronunciation. The other is the "whole-word" theory, which says that people learn to recognize words as a whole and don't bother sounding out the individual letters; only if the string of letters is unfamiliar do they go through the sounding-out process.

Now let's say both of these theories have been simulated by computer models and each one has some success in performing the computer ana-logue of reading words. In order to see which of these models is closer to the way humans actually read, we could design a reaction-time experi-ment to be performed on each of the computer programs, as well as on humans, and see which of the programs produces results which are most in line with the results for humans.

How could we formulate such an experiment? It wouldn't be of any use to ask people to read ordinary words like "cat" or "tender" or "plenti-ful" because these words could be read equally well by both methods and so we wouldn't find out anything by using them. What we have to use are exceptions which can be read by only one of the two methods. Two kinds of exceptions which lead to different predictions by the two theories are words that don't sound the way they are written and letter combinations that can be read but are not words. Examples of the first kind are words like "cough" or "women" or "transition," which are called "exception words." (A wit once suggested that "ghoti" should be pronounced "fish," with the "gh" having the "f" sound it has in "cough," the "o" having the "i" sound it has in "women," and the "ti" having the "sh" sound it has in "transition.") Examples of the second kind are "sint" or "dabinal" or "nulder," which are called "regular nonwords," since they look and sound as though they could have been English words but simply did not happen to be chosen to be words. (To see why these are called "regular nonwords," just compare them to "irregular nonwords," such as "srepchuk" or

"This ghoti isn't ghreti."

"tlathpkwanb," which sound as if they might perhaps belong to some other language, but surely not English.)

If the phonics theory is correct, then people should be able to read regular nonwords more quickly than exception words, since the regular nonwords follow the letter-to-sound route and the exception words don't. If the whole-word theory is correct, however, then people should be able to read exception words more quickly than regular nonwords, since the exception words are recognized as familiar wholes and the regular nonwords are not.

To be sure, even if it is shown that a particular computer model actually works and is also compatible with the way we perform or learn some human function, this does not mean that the model describes the way we "really" do it. There may well be some other model that we haven't thought of yet that is also compatible with what we do, but is closer to what actually goes on inside our heads as we do it. However, this is true of every scientific theory. All we can ever do through our models and experiments is find out whether our model works fairly well or not, or whether one model works better than another; we can never know whether there might not be some even better model out there if only we could think of it.

Semantic network theories

Before the development of connectionist computer models to test theories about mental functions, which began only in the late 1970s, scientists formulated various theories about the functions of the mind that were based much more loosely on the knowledge that the neurons in the brain are organized into networks. The early theories of this sort, which were developed mainly in the 1960s, were known as "semantic network models," and they differed in several crucial ways from the connectionist models I will be describing here. I would like to begin with them, however, for several reasons. First, these early semantic network theories are considerably simpler than the later connectionist models, and they are also more intuitively appealing. Second, some of the connectionist models are actually direct attempts to implement these semantic network theories in computer models. Third, there is an interesting and instructive contrast between the early theories and the more recent models.

In the 1960s, when cognitive psychologists started thinking about how human knowledge is organized, the fact that brain cells are connected in networks gave them the idea that perhaps the knowledge stored in our minds is also organized in networks. They called their theories about these networks of connections in the mind "semantic network" theories because one of the meanings of "semantic" is "related to knowledge of the world." These theories proposed that the various concepts we have – the various things we can think about – are connected to one another in the simplest possible way: Concepts are represented by nodes or points in the network, while the links joining these nodes or points represent the associations between the concepts. For example, "robin" is connected to "bird" because a robin is a kind of bird, and "chick" is connected to "bird" because a chick is a young bird. Moreover, "bird" is connected to "feathers" and "fly" because a bird has feathers and can fly. The links are assumed to be labeled with the name of the relation between the two nodes they connect, so that "robin" is connected to "bird" by an "is a" link going from "robin" to "bird" (because a robin *is a* bird) while "bird" is connected to "feathers" by a "has" link going from "bird" to "feathers" (since all birds *have* feathers) and to "fly" by a "can" link going from "bird" to "fly" (since most birds *can* fly but do not necessarily do so all the time). This simple "bird" network is illustrated in Figure 4.1.

One problem that arises immediately is that not all birds can fly. While it's true that we associate "can fly" with the concept "bird," we also know

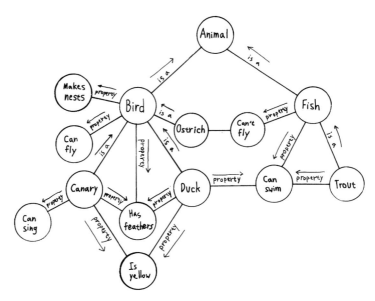

Figure 4.1. Sketch of a portion of a semantic network.

that there are exceptions. When we are asked if birds can fly, we say that most of them do, but not all. The theory accounts for this by adding a node "can't fly" directly linked to those birds that can't fly, such as the ostrich and the emu.

This aspect of the theory illustrates the idea that our storage of facts is "economical." Instead of having a "can fly" node linked to each bird that can fly and a "can't fly" node linked to each bird that can't, there is just one "can fly" node linked to the concept "bird," with a small number of "can't fly" nodes attached to the few exceptions. We can then retrieve the fact that robins can fly by moving along the link from "robin" to "bird" and from "bird" to "can fly."

This "economical" form of representation is used for more general facts as well. For example, since all animals eat, there is no need for a specific link between "robin" and "eats" or even between "bird" and "eats." There is just one link between "animal" and "eats," and if we want to know if robins eat we move along the links from "robin" to "bird" to "animal" to "eats." The process takes a little longer for each extra link in the chain, but eventually we reach the correct answer.

According to the semantic network theory, then, there are links of many different types between the nodes in the network. Some of them connect general concepts with particular examples, some connect items with properties of these items, and some connect items with ideas that are commonly associated with those items, even if they may not be true. You may have read somewhere, for instance, that real ostriches do not actually bury their heads in the sand, but that will not break the link between "ostrich" and "is said to bury head in sand." At most it will add an extra link to the last node, say, "probably not true." This is like the node "can't fly" linked to "ostrich," where our knowledge that ostriches can't fly doesn't eliminate our belief that birds can fly; it just tags this belief as untrue for ostriches.

As we have seen, things that are true of birds in general, such as "can fly" or "has feathers," are usually linked only to the general node "bird," and do not have to be linked specifically to the nodes for particular types of birds, such as "canary" and "penguin." Often, however, some more particular information must be specified for individual birds, such as the shape and color of the feathers. In that case the particular information is linked directly to the particular node – the unique design of peacock feathers to the node "peacock," the bright red color of cardinal feathers to the node "cardinal" – just as "can't fly" is linked to "ostrich."

Labeled links and nodes

Here a very important question arises: If there are so many different kinds of links between the concepts in a semantic network, how does the system distinguish between the links? How does it "know" that "bird" is connected to "fly" by a "can" link, but to "feathers" by a "has" link? The network theorists proposed that each of the links is "labeled" with the name of its type, just as each of the nodes is "labeled" with the name of the concept it represents.

But this presents an even more serious problem. What can there be in the brain that constitutes the labels of the nodes and links in the semantic network? We have seen that the brain consists of neurons and the connections between them; if the neurons are the physical basis of the nodes, and the links the basis of the connections between them, then what could be the physical basis of the labels? As the philosopher Daniel Dennett has pointed out, the information in the network has to be in the system of links itself, not in any labels that could possibly be given to any of the constituents of the network.

To see why labeling the links and nodes would be pointless, we can use

a very old argument known as the "homunculus" argument, which was discovered by ancient Greek philosophers. Our version of this argument goes as follows: For whom do these labels serve as information? There would be no point having labels on the links and nodes in our memory unless there were someone to read these labels. But who could that be? We ourselves cannot read the labels, as they are supposed to be inside us. So there would have to be a little person (a "homunculus") inside us to read these labels. But how would this little person understand what is written on the labels? Inside the homunculus there would have to be another one to read the labels on the links and nodes in the first little person's semantic network. It should be clear by now that this is a never-ending process, and so could not occur in the real world. Thus there is something very wrong with the suggestion that our knowledge of the world is contained in a network of labeled links and nodes.

Another serious problem is that not all of our knowledge consists of associations or relations between two concepts. We also possess a great deal of information about much more complicated relations between larger numbers of concepts. The action "fly," for example, could be associated simply with the object "bird" because "fly" is an intransitive verb. This means that it is something birds do all by themselves, without anything else needing to be involved. But consider the act of giving. How could it be connected to the network? The word "give" certainly has associations for most people. Indeed, it is possible that you are now thinking of the word "present." But the association between "give" and "present" captures only a very small part of the meaning of the word "give." The essence of the word is a relation between four concepts: Someone gives something to someone else. The four concepts involved here are the entity doing the giving, the act of giving, the thing being given, and the entity receiving the thing.

Well, then, why not simply have a network with "give" in the middle, connected with "giver", "receiver" and "object given"? This solution might seem to work for "give," but what would it do for "receive"? "Receive" is connected with the same three concepts as "give," but the relationship is of a different sort. If Carol is giving something to Noel, then Noel is receiving that thing from Carol. Our knowledge about the words "give" and "receive" includes the knowledge that the entity doing the giving in a particular situation is not the one doing the receiving, but rather the one from whom the thing is being received. This complex relationship among the items involved cannot be captured by a semantic network of the type described here.

These and other serious problems with semantic network theories led some of the cognitive psychologists who were studying the organization of knowledge in the mind to look for other theories. A variety of such theories have been proposed, but the one that will occupy us here is connectionist theory, which is described in detail in the next three chapters.

5

What are connectionist networks?

We have seen that there are many problems with semantic networks. The ones we have discussed involve the networks' inability to do many of the things they were designed to do. But there is another, more important reason why semantic networks are inadequate for explaining how the mind works. This is the fact that the human mind has to do much more than just organize information. It also has to perceive new information coming in through the senses and figure out how it fits in with the information already there, and then it has to decide what to do on the basis of all this information. This involves perception and action, and semantic network models were never designed in the first place to be able to handle these human capabilities.

Yet if we look at the brain we see that the areas which process information coming in from our senses, as well as the areas where actions are planned, do not look very different from the areas where various sorts of information are associated or the areas where thinking and reasoning take place. But our basic assumption has been that the way the mind works – as opposed to the actual information it contains – is closely related to the structure of the brain. We therefore ought to look for a theory about the mind's operation that works for perception and action as well as for associations and reasoning.

Transforming semantic networks into connectionist networks

Since we began with semantic networks, let's see if there's anything we can do with them to make them a more plausible candidate for the organizing force of the mind that is based on the way the brain is structured.

Consider the picture of the network of nodes representing concepts connected by links representing the relations between the concepts, as shown in Figure 4.1, in the previous chapter.

Now make the following leap of the imagination: Let the nodes and the links change roles. Try to imagine that the concepts are represented by long chains of links, while it is the relations between the concepts that are represented by the nodes. Think of it this way: The concept "bird," to use our old example, is not a unitary, atomic thing that just happens to be connected to the concepts of "has feathers", "can fly", "robin", "penguin" and a whole host of other concepts. Rather, all the elements associated with the concept "bird" can be seen as constituting this concept. The concept is not merely associated with these elements, it is constructed out of them, and they are an integral part of it. Each concept is thus a whole chain of links.

But if the chains of links form the concepts themselves, then how are the connections between the concepts represented? In our new networks, it is the connections that are represented by the nodes. The chain of links that represents "bird," for example, might be connected at one node to "feathers" and at another node to "nests," as shown in Figure 5.1.

As you can see from the figure, each of the elements constituting the concept "bird" is also a concept in its own right. "Feathers," for example, is not merely part of the concept "bird." It is also a perfectly good concept in its own right, and is connected to other concepts having no direct connection with "bird," such as "pillow" and "quill." Moreover, many of the concepts that constitute the concept "bird" are connected among themselves – "feathers," for example, is obviously connected to "nests."

Thus we again have a network, although of a different sort. It is a much more complex type of network, corresponding to the complexity of our concepts. It is this complex sort of network that is called a "connectionist" network.

The scientists who invented the early models of these networks – and who are still working hard at refining these models, as they began this work only in the 1970s, when they were quite young – are Geoffrey Hinton, James McClelland and David Rumelhart. The three work together with many other researchers in a group called the "PDP Research Group," where "PDP" stands for "parallel distributed processing," their name for their type of connectionist network. As I explain how this sort of network operates, the meanings of the words "parallel" and "distributed" in this context should become clearer.

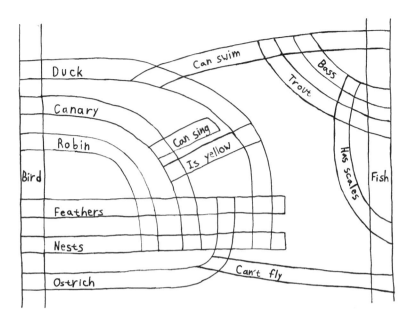

Figure 5.1. Sketch of the general idea of a connectionist network. The "roads" represent the concepts, while the "crossings" represent the links between the concepts. The concepts depicted here probably do not all belong to the same network. Thus some of the links represent connections within the same network, while others represent connections between different networks.

The word "processing" refers to everything the network does, and it is meant to emphasize the most important fact about these networks – namely, that they are dynamic, always working at receiving new information and integrating it with the old information. They are thus very different from most previous models of memory, such as the semantic networks I just described, which are concerned with the way our knowledge is organized, but have very little to say about how it got to be organized that way. Connectionist models are very concerned with how we learn new things, and this will be explained in the next chapter, after we see how these models work.

Another very important point about this network is the fact that, as we have seen, each concept is formed not by a single node but by a whole interconnected set of nodes. But that's what a network is – an interconnected set of nodes. What does this mean, then? Does it mean that every concept is represented by a whole network? If so, where do the other concepts go? How can there be enough room in our brains for the neurons that serve as the basis for all these concepts if each concept takes up a whole network?

Representing concepts by whole networks

The answers to these questions are rather complicated, because the whole system is a very complex one. As a first approximation, the answer to the first question is that every concept is indeed represented by a whole network. This is what is meant by the word "distributed" in PDP – the information in each concept is distributed over a whole network of units. The answer to the second question is that a whole set of concepts of a particular type is located in a single network, and this provides an answer for the third question as well.

However, this answer raises even more questions: How can many different concepts be represented on the same network? How can the network know which concept it is using now if they are all there at the same time? If, as seems most plausible on the basis of the existing evidence, concepts of the same type – animals, such as cat and dog, or colors, such as red and green – are represented on the same network, then how can we react differently to cats and dogs, or to red and green? How can we give them different names if they are all grounded in the same links between the same nodes? How can we love cats and hate dogs, or vice versa? How can we stop at red and go at green?

These questions too have a complicated answer, but I'll try to present it simply. Different concepts can be represented on the same network by virtue of the fact that each concept is represented by a different pattern of activity of the nodes in the network.

Representing concepts by patterns of activity

To understand what a pattern of activity is, let us follow a suggestion put forth by Edward deBono in 1969, some years before any academic research on the topic had begun. DeBono has never been recognized by the academic establishment, perhaps because he publishes his ideas in popular form instead of submitting papers to scholarly journals, but his ideas are the forerunners of the work described in this book.

What deBono asks us to imagine is that a neural network, such as the one in which animals are represented, consists of a square array of tiny light bulbs that can be either off or on, each of these light bulbs being one node of the network. Let's say, he suggested, that there are a hundred of these tiny light bulbs in each row of the square, with a hundred rows of these bulbs. Then "cat" might be represented by the pattern where all the light bulbs in the even-numbered rows are on and all those in the odd-numbered rows are off, while "dog" might be represented by the pattern where all the light bulbs on the left half of the even-numbered rows and the right half of the odd-numbered rows are on, while the remaining light bulbs are off. "Horse," in contrast, might be represented by a pattern where three rows are on, then two are off, then three on, then two off, for the entire array.

It is important to realize that there aren't any little pictures of these animals inside your brain. The analogy of light bulbs is useful because it is vivid and easy to comprehend, but it can create the impression that there are actually picture-like images in the brain. It is therefore important to remember that the nodes are not really light bulbs but neurons, and the states of being on or off are states of these neurons firing or not firing, as described in Chapter 3.

Different kinds of networks

But if the animal-representing network doesn't consist of little pictures of animals, what does it consist of? The answer to this question depends on which one of our animal-representing networks we're talking about, as it turns out that there are several different networks that represent different aspects of animals. There's a network that we might call the animal-image network, which is activated when we see an animal or when we try to picture one in our imagination. Then there's one that could be called the animal-name network, which is activated when we hear the name of an animal or see it written, or when the sight of an animal makes us think of its name. And there's also one that resembles the semantic networks described in the previous chapter, which contains the various concepts associated with each animal and the facts we know about it, and so we might call it the animal-concept network. The reason we believe all these networks are separate is that it's possible for any one of them to be damaged while the rest of them remain intact.

Well, then, you might ask, doesn't the animal-image network, at least, contain little pictures of dogs and cats, say? What else could be in it? How can it help us see dogs and cats if it doesn't have little pictures of dogs and cats in it? The "homunculus" argument I presented in the previous chapter shows why this is impossible. There we saw that there would not be any point in having labels on the links in the networks, as there would have to be some little person inside our brains, looking at these labels. In the same way, there would be no point in having pictures in the networks, as once again there would have to be another little person inside this one's brain to look at them, and so on ad infinitum. As in the case of the labeled links, this is clearly impossible, so there has to be some other way that we form our images of cats and dogs.

Connections between the networks

What is this other way, then? If we don't have little pictures stored in our brains, how can we know what we're looking at? Well, first the retinas in our eyes are activated by the light from the object in front of us, and then the neurons in the retina activate neurons in the primary visual centers of our brain, which respond to such things as lines and curves, or light and shade, or red and green. But then how do we know whether it's a cat or a dog, or a tomato or a pepper, that we are looking at?

Well, that depends on what you mean by knowing it's a cat or a dog. If you mean knowing its name, the way we know it is that the animal-name network is connected to the animal-image network, which is connected to the primary visual areas of our brain. When we see a dog, say, the pattern created in the visual areas of our brain, which is composed of the lines and curves and colors and ways of moving characteristic of dogs, activates the neurons in the animal-image network that have learned to fire when this pattern occurs. (How the neurons in one net learn to fire in a particular pattern when activated by a particular pattern in another net is discussed in the next chapter.) This pattern of activation in the animal-image network then produces the pattern of activation associated with dogs in the animal-name network, which enables us to think, "There's a dog."

But you might be thinking of something else when you consider what it means to know that the animal you're looking at is a dog. Maybe you're thinking of knowing various facts about it, such as the fact that it can bark, or that it can bite, or that it can be a "man's best friend." Well, as we've already seen, facts about what dogs can do are part of our animal-concept network, which is connected directly to both the animal-image network and the animal-name one. Thus the pattern of activation produced by seeing a dog can directly activate the facts we know about dogs.

But there can also be other associations involving dogs. Hearing the

word "dog" can lead us to think about what dogs do just as easily as the
sight of a dog, as there are direct connections between the animal-name
network and the animal-concept network as well. Moreover, hearing the
word "dog" can also activate common phrases containing the word "dog,"
such as "raining cats and dogs," which are unlikely to be activated by the
sight of a dog. Such phrases may well form part of a different sort of
network, an "animal-phrase" network, which is likely to be directly connected to the animal-name network but not to the animal-image
network. The animal-phrase network is discussed in Chapter 7, where the
relations between concepts are explored.

Two-way connections between the networks

Moreover, these processes are not merely one-way. Although the activation of one neuron by another is usually one-way, as we saw in Chapter 3,
the use of different neurons can make it possible for many networks to be
connected in both directions. For example, while some connections
between the animal-image and the animal-name networks allow the
image of a dog to activate the word "dog," other connections allow the
word "dog" to activate the image of a dog. But then how do we know that
we are not actually seeing a dog when we form an image of a dog in our
animal-image network?

The answer to this question is not very clear at present, but one possibility is that all the various networks that have just been activated retain
some of their activation for a while. Our knowledge is based not only on
the patterns of activated neurons within a particular network but also on
the pattern of all the networks activated at a given time. Thus the total
state of our brain, and therefore of our mind, when the animal-image
network is activated "from within" – that is, by other networks in our
brain – is different from the state when the same network is activated by
the actual sight of a dog.

Cooperation among the networks

We have just seen that connections between networks can be two-way.
But in the previous section we saw that any particular network can activate more than one other network, as when the animal-image network
activates both the animal-name network and the animal-concept
network. Putting the two types of process together, we can get a third

type, in which activation in two different networks at the same time activates a third network to which both of them are connected.

How could this work? What sort of phenomena could it explain? Well, let's say someone showed you a picture of an unfamiliar animal that looked somewhat like a leopard but not quite. You know it's some sort of big cat but you can't think of its name. Then the examiner gives you a hint: The animal's name starts with the letter "o." Right away the task becomes much easier, and you say "ocelot."

What is happening here? When you see the picture, various names of big cats are activated in your animal-name network. The activations of such names as "tiger" and "leopard" are fairly high, but some details of the picture convince you that the animal is neither a tiger nor a leopard. This provides some inhibition as well for these two names, which lowers their activation to the point where you do not respond with "tiger" or "leopard." These names nevertheless remain too highly activated to allow the network to switch to the pattern for "ocelot," which is more weakly connected because it is less frequently used.

But then you hear the name of the letter "o," which activates another network, containing the letters of the English alphabet. This is a very interesting network because it is position-sensitive. Its patterns do not represent merely the individual letters; there are different representations for, say, "g" as the first letter of a word and "g" as the last letter. This network is clearly strongly connected with all your word networks, and it has both activating and inhibiting effects. In the case of the letter "o," for example, it provides extra activation for words that begin with "o" and inhibits the patterns for words that do not begin with "o."

Now the pattern for the letter "o" as a first letter is connected to only a few words in the animal-name network, but, still, on its own it might not activate "ocelot" in this network, since "ox" and "owl" are more common. But recall that you are also looking at a picture of an ocelot, and this picture has specifically provided some activation for all the names of the big cats in your animal-name network. The activation of words beginning with "o" coming from the letter network, together with the inhibition of words not beginning with that letter, thus combines with the activation of big cats coming from the animal-image network to boost the activation of the name of a big cat beginning with the letter "o." This allows the pattern for "ocelot" to dominate the animal-name network, and you say "ocelot."

Overlapping representations

But here we are faced with a fundamental dilemma. We have seen that the images of different animals, say dogs and cats, are represented by different patterns of activity in the animal-image network, while the words "dog" and "cat" are represented by different patterns of activity in the animal-name network. However, we also know that the neurons in the animal-image network are connected with the neurons in the animal-name network by a single set of connections. Then how can the image of a dog activate the word "dog" rather than the word "cat," and how can the image of a cat activate the word "cat" rather than the word "dog"? Why doesn't the sight of a dog activate all the neurons in the animal-name network rather than just those that represent the word "dog"?

The answer to this question lies in the fact that the connections between the neurons in the two networks do not all have the same strength – some are stronger than others, and there is a great deal of variation in the strengths of the different links. In Chapter 3 we saw that the strengths of the connections between neurons can vary as a result of differences in the various stages in the transmission of messages from the first neuron to the second. To understand how the differences in the strengths of these connections make it possible to convey different messages across the same set of links, it is not necessary to know what led to these differences. All we need to know is that these links can be stronger or weaker. In fact, connectionist models simply use numbers from -1 to $+1$ to represent the strengths of the connections between the nodes. For example, $+1$ means that the first unit activates the second very strongly, $+0.5$ represents a moderate activation of the second unit by the first, 0 means no activation, -0.5 represents a moderate inhibition of the second unit by the first, and -1 means that the first unit inhibits the second very strongly.

The fact that the neurons in the animal-image network (let's call it AIN) are connected to the neurons in the animal-name network (call it ANN) means that a typical neuron in ANN can receive input from tens of thousands of neurons in AIN. What it then has to do is add up all these inputs and "decide" whether or not to fire. This "decision" is based on the threshold of activation of that particular neuron. That is, each neuron has a particular number that is the minimum input it needs in order to fire. If the total input it receives is greater than this threshold number, then it fires; if not, it does not fire. As we have seen, each of these inputs can be represented by a number between -1 and $+1$.

Now if the dog image is activated, some of the neurons in AIN will be firing while others will be silent. A neuron in ANN will thus receive input only from those neurons in AIN that are firing at the time. It will then add up the inputs from these neurons, and it will fire only if the total is greater than its threshold for firing. As this happens for each neuron in ANN, some of them will fire while others will not, thus creating a new pattern of activity in ANN. This pattern represents the word "dog."

Actually working through an example of how this takes place will make it much clearer. Since an example consisting of ten thousand units in one network linking up with ten thousand in another would be impossible to follow – and practically impossible to construct – we will use a simplified example of two networks with six units each. This example has numbers in it, but the only mathematics it requires is the ability to add a column of up to six numbers, and even that can be done on a pocket calculator, so try to follow along. It's really important, because it's the very heart of connectionist theory. You can take my word for it if you want, but it's actually very simple if you take it step by step.

Let's say, then, that our dog image is the pattern 011001 on this six-unit network. This means that neurons 2, 3 and 6 in AIN are firing while neurons 1, 4 and 5 are silent. Our cat image, say, is 110010, which means that neurons 1, 2 and 5 are firing, while 3, 4 and 6 are silent. You can see that neuron 2 is activated in both the dog and the cat images; in the real case, there will be many neurons that are activated in both images, reflecting the fact that there are many aspects in which dogs and cats are similar.

Now let's consider how these six neurons are connected with neuron 1 of ANN, which has, say, a threshold of 0.5. Let's say that the links have the strengths 0.1, 0.2, 0.3, 0.2, 0.1 and 0.2. Then we can set up the chart shown in Table 5.1.

Table 5.1

Dog image	0	1	1	0	0	1
Cat image	1	1	0	0	1	0
Links to ANN 1	0.1	0.2	0.3	0.2	0.1	0.2

Now we have to figure out the inputs of each of the AIN neurons to neuron 1 of ANN, and then add them up to see if the sum passes its threshold of 0.5. Consider the dog image first. Its first number is 0, which means that neuron 1 of AIN is not activated, so we can ignore its connection strength. The second number is 1, which means that neuron 2 is activated, so we add its strength of 0.2. As the third number is also 1, neuron 3 is also activated, so we further add its strength of 0.3, for a total of 0.5. Now the next two numbers are 0, so neurons 4 and 5 are both silent, and we ignore their connection strengths. The last number is again 1, so we add in the connection strength of neuron 6, which is 0.2, for a grand total of 0.7. This passes the threshold of 0.5 that we postulated for neuron 1 of ANN, so this neuron is activated whenever the dog-image pattern is activated. Thus the first number of the dog-name pattern is 1.

Now let's consider the links to the other five neurons in ANN, as well as the thresholds of these neurons, and add them all to our chart to find out which of these neurons are firing when the dog-image pattern is activated in AIN. This is shown in Table 5.2. (All the numbers are arbitrary, made up for the purpose of providing an instructive example.)

Table 5.2

	AIN 1	AIN 2	AIN 3	AIN 4	AIN 5	AIN 6
Dog image	0	1	1	0	0	1
Links to ANN 1	0.1	0.2	0.3	0.2	0.1	0.2
Links to ANN 2	0.2	0.1	0.1	0.1	0.1	0.2
Links to ANN 3	0.3	0.2	0.1	0.1	0.2	0.1
Links to ANN 4	0.3	0.1	0.2	0.1	0.1	0.1
Links to ANN 5	0.1	0.2	0.2	0.1	0.1	0.2
Links to ANN 6	0.3	0.3	0.1	0.1	0.2	0.2

We can figure out the inputs to the other five neurons in the ANN the same way we did for the first one, by adding up the strengths of the links from neurons 2, 3 and 6 and ignoring the others, since only neurons 2, 3 and 6 are firing when the dog-image pattern is activated. Table 5.3 shows us what we get.

Table 5.3

	Link 2	+Link 3	+Link 6	Sum
ANN 1	0.2	+0.3	+0.2	0.7
ANN 2	0.1	+0.1	+0.2	0.4
ANN 3	0.2	+0.1	+0.1	0.4
ANN 4	0.1	+0.2	+0.1	0.4
ANN 5	0.2	+0.2	+0.2	0.6
ANN 6	0.3	+0.1	+0.2	0.6

Now we compare each sum with the activation threshold for that neuron in the ANN, to see if it is activated when the dog-image pattern is firing (see Table 5.4). (Once again, the numbers for the thresholds are made up for the example.)

Table 5.4

	Sum	Threshold	Yes (1)/No (0)
ANN 1	0.7	0.5	1
ANN 2	0.4	0.6	0
ANN 3	0.4	0.6	0
ANN 4	0.4	0.4	1
ANN 5	0.6	0.6	1
ANN 6	0.6	0.5	1

Thus the animal-name network acquires the pattern 100111 when the dog-image pattern is activated. This, then, is the pattern of activity for the word "dog."

Now we have to see how the same links between the same two groups of six neurons can yield a different pattern of activity for the word "cat" when the cat-image pattern is activated in the animal-image network. Table 5.5 is a copy of Table 5.2, which lists the dog-image pattern and the links between the neurons, with the dog-image pattern in Table 5.2 replaced by the cat-image one in Table 5.5.

Table 5.5

	AIN 1	AIN 2	AIN 3	AIN 4	AIN 5	AIN 6
Cat image	1	1	0	0	1	0
Links to ANN 1	0.1	0.2	0.3	0.2	0.1	0.2
Links to ANN 2	0.2	0.1	0.1	0.1	0.1	0.2
Links to ANN 3	0.3	0.2	0.1	0.1	0.2	0.1
Links to ANN 4	0.3	0.1	0.2	0.1	0.1	0.1
Links to ANN 5	0.1	0.2	0.2	0.1	0.1	0.2
Links to ANN 6	0.3	0.3	0.1	0.1	0.2	0.2

The inputs to the neurons in ANN are calculated in the same way as for the dog-image pattern. In the case of the cat-image pattern we add up the strengths of the links from neurons 1, 2 and 5 and ignore 3, 4 and 6, since only neurons 1, 2 and 5 are firing when the cat-image pattern is activated. Table 5.6 (overleaf) shows us what we get.

Table 5.6

	Link 1	+Link 2	+Link 5	Sum
ANN 1	0.1	+0.2	+0.1	0.4
ANN 2	0.2	+0.1	+0.1	0.4
ANN 3	0.3	+0.2	+0.2	0.7
ANN 4	0.3	+0.1	+0.1	0.5
ANN 5	0.1	+0.2	+0.1	0.4
ANN 6	0.3	+0.3	+0.2	0.8

Here too we compare each sum with the activation threshold for that neuron in the ANN to see if it is activated when the cat-image pattern is firing. This is shown in Table 5.7.

Table 5.7

	Sum	Threshold	Yes (1)/No (0)
ANN 1	0.4	0.5	0
ANN 2	0.4	0.6	0
ANN 3	0.7	0.6	1
ANN 4	0.5	0.4	1
ANN 5	0.4	0.6	0
ANN 6	0.8	0.5	1

Thus the animal-name network fires in the pattern 001101 when the cat-image pattern is activated. This, then, is the pattern of activity for the word "cat."

Together, these two calculations show how the same links between the neurons of two networks enable two different patterns of activity on the first network to bring about two different patterns on the second one. Here, as mentioned, I set the strengths of the links and the thresholds arbitrarily to achieve the desired result. The next question therefore is how this is achieved naturally – that is, how the links between networks change as a result of experience to form the patterns of activity that represent all the different things we know. But change as a result of experience is simply the definition of learning, so what we are asking is how our networks learn. This is the topic of the next chapter.

6

How our networks learn

How could the sort of connectionist networks described in the previous chapter come into being? How can the neurons in our animal-image network, say, become connected up in just the right way to the neurons in our animal-name network, so that the pattern of activity produced in the animal-image net by the sight of a dog leads to a pattern of activity in the animal-name net that makes us think of the word "dog"?

If each concept were represented by a single node, as in the old semantic network theories, the answer to this question would seem to be simple. It seems as if all we would have to say is that the node representing the image of a dog is connected with the node representing the word "dog." But behind this apparent simplicity lies a great difficulty. How do the two nodes become connected? What force could link up the dog-image node precisely with the dog-name node and not with any other node? Just as there cannot be any "little person" inside looking at little pictures in our brains, there also cannot be any "little person" linking up the proper nodes. Then how are they linked up? The single-node theory may seem to provide a simpler description of the connections between concepts, but it is very difficult to explain learning in such a system.

In connectionist networks, however, as we saw in the previous chapter, dog images and cat images are different patterns of activity on the animal-image network, while the words "dog" and "cat" are different patterns of activity on the animal-name network. The description of how the two networks are connected so that the dog-image pattern gives rise to the dog-name pattern, while the cat-image pattern gives rise to the cat-name pattern, was indeed rather complicated. One of the advantages of connectionist theories, though, is that it is not much more complicated to describe how two such patterns are connected up in the first place – in other words, how learning occurs – than it is to explain what the connections consist of.

From random to structured neural activity

To see how learning occurs, let us first consider what kind of neural activity takes place in the brain of a child who has not yet learned the words "dog" and "cat." All parts of the brain are active from the time that they are formed, so there is neural activity in the networks intended for language even before a human child learns to speak. The difference between the brain of a child who has not yet learned to talk and one who has is that the neural activity in the language areas of the brain of the nontalking child is random, while that of the talking child is structured. It is this random activity in the infant's brain that produces the babbling which is much the same for all babies before it is modified by the baby's linguistic environment. The question, then, is not how activity is created in a non-active network, but how structured patterns of activity are formed in a network in which the neurons have been firing randomly.

At the outset, the neurons in the network which is to become the animal-image network are connected with the neurons in the one which is to become the animal-name network. Learning takes place by a process in which some of these connections become stronger, while others become weaker. We saw in Chapter 5 that connectionist theories are based on the idea that different connections between neurons have different strengths, and that it is these strengths that determine which concepts are associated with which other concepts. Learning thus consists of changes in the strength of the connections.

Learning to say "dog"

Let's see how learning can be described in these terms. We start out with two networks in one-year-old Sarah's brain that are connected in both directions – that is, there are connections through which the neurons in the first network activate those in the second network, and other connections through which the neurons in the second network activate those in the first one. Now let's say Sarah is taken out for a walk every day in an area where there are lots of dogs but no cats. Sarah's attention is caught by a brown dog, and her father, seeing her pointing excitedly at the dog, says, "Yes, Sarah, there's a dog." Then a black-and-white dog runs by, and the father says, "There's another dog." This continues on and off for a couple of months, and one day, while Sarah's father is thinking about something else, a dog runs by and he doesn't say anything, whereupon Sarah yells, "Dog!" What has happened in Sarah's brain?

What has occurred is something like this: The sight of the brown dog has caused certain specific neurons to fire in Sarah's visual center – neurons associated with shape, color and movement. Since these neurons are connected to the ones in the part of Sarah's brain that is destined to become the animal-image network, they activate the neurons there with which they happen to be connected most strongly on the basis of the way they were wired up as the brain developed. The particular neurons that are activated in the future animal-image network form the pattern of activity that constitutes the visual image of this brown dog as seen running from the side.

Now one of the most important principles of learning is Hebb's law, which was mentioned in Chapter 3: If a neuron fires just after it receives an input from another one, the connection between them is strengthened, so that the receiving neuron becomes more likely to fire again the next time it is activated by the same neuron. This is a tiny increase in the strength of the connection, but if it takes place many times it can lead to a very strong link between the two neurons. What this means in our present case is that the same neurons in the animal-image network will fire again the next time Sarah sees this brown dog in the same perspective.

Then what happens when Sarah sees the black-and-white dog – or even if she sees the brown dog again, but from a different angle? The two dogs, or the two perspectives of the same dog, are fairly similar to one another – much more similar, say, than either of them is to a table or a house or a car. Therefore the sight of them will activate many of the same neurons in the visual center, and these in turn will activate many of the same neurons in the animal-image network. But the connections between the neurons in the visual center and the ones in the animal-image network that were activated by the sight of the brown dog from the side are all now a little stronger than they were before. Thus when the neurons in the visual center are activated by the sight of the black-and-white dog, the ones that overlap with those that were activated by the sight of the brown dog will fire more strongly. As a result, the pattern of activity in the animal-image network that is associated with the black-and-white dog is likely to be even more similar to the pattern of activity associated with the brown dog than the second dog is to the first one.

But what does it mean to say that one pattern of activity is similar to another? How does one measure similarity in patterns of activity in neural networks? We saw in Chapter 5 that each pattern of activity consists of an array of neurons, some of which are firing and others of which are not firing at that particular time, so that each such pattern can be represented by an array of o's and 1's. To use our six-node example, the array 001101 is considered more similar to 000101 than it is to 100101 because the value of only one unit has to be changed to go from 001101 to 000101 while the values of two units have to be changed to go from 001101 to 100101. In general, then, two patterns of activity are more similar to each other the more of their units have the same value.

Now let's get back to Sarah. At the same time that she sees each of the dogs, she also hears her father say "dog" – this is the stressed word in his sentence, and so it is the one that makes the strongest impression. Hearing this word activates certain neurons in Sarah's language-hearing center, which are connected to the part of her brain that is to become the animal-name network. As in the case of the visual image, these neurons then activate those neurons in the future animal-name network that they happen to be connected with most strongly at the time. The pattern of activity formed by these neurons is now associated with the word "dog."

But this is not all. Remember that the two inner networks – the animal-image one and the animal-name one – are also connected with each other. Moreover, the neurons that form the pattern for the image of the dog Sarah is looking at are firing at about the same time as the neurons forming the pattern for the word "dog." Therefore the Hebb learning principle operates here as well, and the links between the neurons in the two patterns of activity are strengthened. As before, these connections are strengthened only a little at a time, but with repeated occurrences of seeing dogs at the same time as hearing the word "dog," they become quite strong. As a result, each pattern becomes able to activate the other by the process described in Chapter 5. Thus after Sarah has heard "dog" while looking at dogs a number of times, the auditory image of the name is strongly activated the next time she sees a dog, even if the word is not being spoken at the time. Other networks that are linked in a similar fashion eventually activate Sarah's vocal apparatus, and she says "Dog!"

Classifying things in the same category

But how is it that Sarah learns to say the same word "Dog!" both when she sees the brown dog and when she sees the black-and-white dog? Doesn't each pattern in the animal-image network activate a different pattern in the animal-name network, as we saw in Chapter 5? Well, sometimes it does and sometimes it doesn't. In fact, the problem of when it does and when it doesn't is connected with one of the most interesting issues in the study of the mind, the question of how we classify objects into categories. When two images evoke the same pattern on a name network, we say that the two objects have been classified as belonging to the same category; when they evoke two different patterns, we say they have been classified in two different categories. Let's see how these processes occur.

First we must remember that the patterns of activity in the animal-image network for the brown dog and the black-and-white dog are very similar to one another. In fact, let's assume that they are more similar to each other than either of them is to any other pattern in that network. Then when the black-and-white-dog pattern is firing, it activates most of the same neurons in the animal-name network that are activated by the brown-dog pattern.

Here we have to consider another crucial aspect of the way these networks operate, one similar in importance to the Hebb learning principle. This is the fact that a pattern in a given network can be activated not only by a pattern in another network but also by another pattern in the very same network. Such within-network activation is made possible by the fact that the neurons within a network are all connected with one another – some directly, some indirectly – as we saw in Chapter 3. Here again there is a habituation principle at work: If a pattern in a given network, say the animal-name network, has been activated a number of times, then it tends to be activated again.

A pattern of this sort is called an *attractor*, because other patterns tend to activate it, while it continues to activate itself and does not lead to the activation of other patterns. It is as if other patterns are "attracted" to it and cannot get away. Specifically, the attractor pattern is activated by other patterns that are very similar to it, or at least more similar to it than to any other pattern. Each time the attractor is activated, the links connecting the other patterns to it are strengthened, according to the same learning principle as the one that operates between networks. This makes it even stronger, which means that patterns similar to it don't become attractors themselves, since they always lead to the attractor that already exists.

What does this tell us about Sarah and the dogs? Well, if her brain has already made the connection between the brown-dog image and the word "dog," and the pattern for the word "dog" is one of the few familiar patterns in Sarah's animal-name network, then when she sees the black-and-white dog again it activates almost the same neurons as the ones activated by the brown-dog image. But the pattern activated in the animal-name network by the brown-dog image is the one associated with the word "dog." Thus the pattern activated in this network by the black-and-white dog image is very similar to the dog-name pattern, and since the dog-name pattern is an attractor, it is immediately activated by this new pattern. Thus the links between the units of the black-and-white dog-image pattern and the dog-name pattern that are the same as the ones between the brown-dog pattern and the dog-name pattern are strengthened, while the links to other units in the animal-name network are gradually weakened. As a result of this process, the dog-name pattern will be activated whenever Sarah sees the black-and-white dog.

Classifying things in different categories

Now suppose that at the same time that Sarah's father has been pointing out dogs to her and saying, "There's a dog," he has also been pointing out birds flying in the sky and alighting on trees and saying, "There's a bird." Since the appearances and actions of the dogs and the birds are very different, they form patterns of activity on the animal-image network that differ by many units. Therefore they also activate very different patterns on the animal-name network, so that neither of them serves as an attractor for the other. Sarah will thus learn to classify dogs and birds in separate categories, and she will have no trouble calling them by their accepted names.

But let's say Sarah's father takes her for a walk one day in a new neighborhood where there are cats as well as dogs. Sarah sees a cat running down the street, and she says "Dog!" Why does she do this?

Actually, given the way the networks operate, it would be surprising if she didn't. Clearly, the mechanism at work when Sarah says "Dog!" upon seeing a cat is exactly the same as the one at work when she says "Dog!" at the sight of the black-and-white dog after having heard "There's a dog" when she was looking at a brown dog. Since Sarah has not yet learned the

"You've got to stop overgeneralizing all the time."

word "cat," the attractor most similar to the pattern activated by the cat image is the dog-name pattern, and so this attractor pattern is quickly activated, causing Sarah to say "Dog!"

Indeed, the only difference between the two cases is in our expectations. We expect children to say "Dog!" at the sight of a new dog after having heard the word in connection with other dogs, so we consider this an acceptable generalization – in fact, it seems so natural to us that we hardly notice it at all. On the other hand, when Sarah says "Dog!" at the sight of a cat it goes against our usual way of talking about cats, so we say she has made a mistake and we correct her.

People who study child development often refer to the child's calling a cat "dog" as "overgeneralization," as opposed to "correct generalization." These people seem to be assuming that two different processes are involved, but actually there is only one. To be sure, we have to teach children to use the same words as everyone else in order to communicate with others, but they are not making a mistake when they use a word they already know to refer to something similar that we happen to call by some other name. They are engaging in the all-important process of generalization, which is a cornerstone of human thinking.

One way that we can realize the importance of this process is to look at what happens in the rare cases when it is missing. People with autism often suffer from the lack of the ability to generalize. One of the early signs of autism may be that a child will use a word correctly once and then not use it again, perhaps for years. It has been speculated that the autistic child associates the word with the exact view of that exact object and cannot generalize it to the sight of other objects of the same sort or even different views of the same object. In network terms, this child does not have any patterns that function as attractors. Therefore such a child will not use the same word when confronted with a slightly different object, and, more importantly, will be unable to apply the facts he has learned about one object to other objects of the same type.

How generalization works

Let's take a closer look now at the neural mechanism that underlies the normal child's ability to generalize, whether "correctly" or "incorrectly." Consider the patterns of activity on Sarah's animal-image network. In Chapter 5 we used six-unit patterns to represent the links between patterns in the animal-image and animal-name networks in order to simplify the calculations, but six units do not allow us enough room to maneuver in an attempt to explain how generalization occurs. Let's try ten-unit patterns here, keeping in mind that the actual patterns in our brain consist of tens of thousands of neurons.

Let's say, then, that the brown dog Sarah originally saw is represented by the pattern 0011010101 on the animal-image network, while the black-and-white dog she sees later is represented as 0011000101 (a difference in the value of only one node, the sixth one). Through the process of habituation described earlier, the brown-dog-image pattern 0011010101 has become able to activate a dog-name pattern in the animal-name network – say, 0110000111. Now when Sarah sees the black-and-white dog, the neurons that are activated in the animal-name network are likely to be mostly the same as the ones activated by the sight of the brown dog. But we recall the principle that when a network is in some new, unusual pattern of activity, it tends to activate the attractor pattern to which the new pattern is most similar. Since the pattern in the animal-name network activated by the image of the black-and-white dog is an unusual one, while the more frequently occurring pattern for the brown dog is very similar to it, the animal-name network will quickly slide into the pattern of activity associated with the brown dog, and Sarah will say "Dog!"

Now suppose that Sarah's bird image is the pattern 1100111000 (differing from the brown-dog and the black-and-white-dog patterns by eight and nine units, respectively). And since she has also learned the word "bird," let's say the bird-image pattern activates the pattern 1000101011 on the animal-name network. What happens, then, when Sarah, equipped with these links between her animal-image and animal-name networks, now encounters a black cat?

Let's say the cat image is represented by the pattern 1011000111 (differing from that of the brown dog by three units and that of the black-and-white dog by two, but differing from the bird-image pattern by nine units). According to our criterion for similarity between patterns of activity, the cat-image pattern is much more similar to either dog-image pattern than it is to the bird-image pattern. Therefore the pattern it activates in the animal-name network is very similar to the dog-name pattern but very different from the bird-name pattern. The dog-name pattern will thus "attract" it – in other words, the pattern for the word "dog" will quickly be activated, and Sarah will say "Dog!"

How differentiation works

Now that we understand why Sarah says "Dog!" when she first sees a cat, how can we explain the fact that she does eventually learn to say "Cat!" on later occasions? Well, there are two opposing forces at work here. On the one hand, the cat image evokes the dog-name pattern on the animal-name network, which tends to strengthen this pattern even further. On the other hand, hearing her father say, "There's a cat!" when she sees a cat evokes a new pattern on the same network, a pattern representing the word "cat." At first the older pattern is more strongly evoked than the new one, so Sarah continues to say "Dog!" for a while when she sees cats. But as her father persists in saying, "No, that's a cat" on each occasion, the links leading to the new cat-name pattern are gradually strengthened, and eventually they become strong enough to cause Sarah to say "Cat!" when she sees a cat.

These changes in the links are a slow and complex process, because they have to keep the connection between the dog-image pattern and the old dog-name pattern while changing enough to link the cat-image pattern with the new cat-name pattern. The change is accomplished by very small successive adjustments to the strengths of the links between the two networks, until the patterns become differentiated, and Sarah is able to say "Dog!" when she sees a dog and "Cat!" when she sees a cat.

This explains a very general phenomenon that occurs with all of us, adults as well as children. Whenever we encounter a new class of things, we tend to see the members of this class as being all alike. If we are

Lawyer: "Dr. Chan, which police officer gave you the sample for testing?"
Expert witness: "Sorry, Ms. Jones, I can't remember. All you Westerners look alike to me."

Westerners who have had very little contact with Asians, we say that all Orientals look alike – we cannot distinguish between Chinese and Japanese people, or between one individual Chinese person and another. The same thing occurs in the other direction as well – if we are Asian people who know very few Westerners, we cannot distinguish between English and Italian people, or between one individual Italian person and another.

This is due to the same process as the one that makes a child say "Dog!" upon seeing a cat. The first object seen in the new class forms an attractor in the appropriate network, and so other items of this class tend to activate it. It is only after a period of learning and becoming more closely acquainted with various members of the new class that our brain is able to make the small adjustments in the links that lead to the formation of many new patterns to represent all these different people or things.

When is learning satisfying and effective?

This process of making small adjustments to form new patterns that represent new information can also provide an explanation for a very interesting aspect of learning in both children and adults that was discovered by the Russian psychologist Vygotsky early in the twentieth century.

Vygotsky developed a theory called ZPD, or zone of proximal development. The essence of this theory is that people attend with greatest interest and learn the most when they are presented with ideas that are just a little bit beyond what they already know. We all know how bored we can get when we are forced to listen to things we already know very well, and how quickly we feel lost when presented with a barrage of totally new information.

How can this phenomenon be explained? If learning were simply a matter of accumulating lists of facts, then it shouldn't make any difference if we are presented with information that is just a little bit beyond what we already know or totally new information. Each fact would simply be stored separately. According to connectionist theory, however, our knowledge is organized into patterns of activity, and each time we learn something new we have to modify the old patterns so as to keep the old material while adding the new information. The adjustments are clearly smallest when the new information is only slightly new – when it is compatible with what we already know, so that the old patterns need only a little bit of adjustment to accommodate the new knowledge. If we are trying to understand something totally new, however, we need to make larger adjustments to the units of the patterns we already have, which requires changing the strengths of large numbers of connections in our brain, and this is a difficult, tiring process.

This idea also explains why we understand a new theory more easily and feel a greater sense of satisfaction when it is described in terms of an old one. A well-known example of this phenomenon is the analogy between the structure of the atom and the structure of the solar system. In this case we already have a network representing the structure of the solar system, so we don't have to create a whole new network – we can take the old one and modify it slightly by connecting the various parts of the visual structure with the new names appropriate to the new theory. When we learn the differences between the two systems, this again requires only small modifications of the old system, something we find much easier than creating a whole new system.

Since learning totally new things is so difficult, we are always looking for analogies between the various things we know. Our desire for analogies is so strong that we often see them where they do not exist, and anyone who can create new analogies for us is greatly admired. This may be one of the reasons why people enjoy metaphors so much. A great deal of study has gone into explaining how metaphors are understood – how we understand what the poet is trying to say when he or she says one thing in terms of another – but one puzzle that is rarely addressed is why anyone should bother saying one thing in terms of another in the first place. Why not just say what you mean directly?

Once we see that understanding one thing in terms of another is fundamental to the learning process, we can see why we so greatly admire those people who can show us likenesses that we never saw before. As Aristotle said long ago, metaphors are admired because they give us new knowledge. Connectionist theory explains why this sort of new knowledge is often so much more satisfying than other sorts.

7

Connecting the networks: how different things are related

Word-association tests notoriously reveal that people have a tendency to say "dog" when the tester says "cat", "low" when the tester says "high", "potatoes" when the tester says "meat," and "mother" or "son" when the tester says "father." Yet the types of relation between these pairs of words are very different. Cats and dogs are both animals – members of the same category – just as meat and potatoes are both foods, and lions and tigers are both wild cats. Each pair is mutually exclusive, in the sense that no animal can be both a dog and a cat at the same time; if an animal is a cat, it is not a dog, and vice versa.

High and low, in contrast, are opposites. They are generally thought of as being at the ends of a continuum, but they are not mutually exclusive in the same way that dogs and cats are. That is, the same thing can be considered either high or low, depending on the context. For example, we may call a certain mountain high when we are speaking about it in the context of the other mountains in its vicinity that are lower than it, yet if we are speaking about the highest mountains in the world, we may call the same mountain low.

The relationship between parent and child is yet a different one. These are opposites in a different sense – if Leah is the mother of Simon, then Simon is the son of Leah. The relationship between Leah and Simon is directly linked to the one between Simon and Leah, but it is not the same one.

Thus the way the concepts of "mother" and "son" are connected in our mind must be somehow different from the way "cat" and "dog" are connected, and different yet again from the way "high" and "low" are connected. This raises the question of how a theory such as the present one, which seems to be trying to explain all our knowledge on the basis of associations, can handle the fact that there are different kinds of links between different words that are associated with each other. In Chapter 4 I discussed the explanation offered by the semantic network theorists, which put different labels on the different types of links. I showed that this explanation is unacceptable in a theory that is modeled on the structure of the brain, because there is no place for labels on the connections between groups of neurons in the brain. But then how can we explain the different types of relation?

Fact networks

Let us recall that concepts of the same kind are represented by patterns of activity of the neurons in a particular module, with each different kind of concept represented in a different module. Thus "cat" and "dog" are both in the animal module, while "meat" and "potatoes" are in the food module, and "mother" and "son" are in the family-members module. Now let's make a leap: Geoffrey Hinton, a leading figure in connectionist theory, has suggested that not only are concepts represented by patterns of activity in neuronal networks, but so are the facts that connect these concepts. What this means is that there is, for example, an animal-fact network where the patterns of activity represent facts such as "Cats are animals" and "Dogs are animals," just as the animal-name network has "cat" and "dog" patterns. This module takes over the function of the labeled links in the semantic network connecting "cat" with "animal" and "dog" with "animal."

But how does this work? Let's say you are participating in an experiment and you are asked, "Is a cat an animal?" The mention of "cat" activates the "cat" pattern in your animal-name network, which in turn activates not only the "dog" pattern in the same network and the "meow" pattern in your animal-sound network, but also the "Cats are animals" pattern in your animal-fact network. Now the question "Is a cat an animal?" shares its two main elements, "cat" and "animal," with the sentence "Cats are animals," and so it causes the strongest activation of that particular sentence in your animal-fact network.

At the same time, you also know the general relation between sentences that are questions and sentences that are the answers to the questions. This relation is a grammatical transformation between "An X is a Y" and "Is an X a Y?", which connectionists believe must be stored in yet another type of network, although we do not yet know how this type of network operates. The transformation relation enables you to answer "Yes" to the question that has been posed. You have no need for labeled links between "cat" and "animal" because you have a network that is connected to the animal-name network and supports patterns of activity that represent the relation between "cat" and "animal" as the sentence "Cats are animals."

Distinguishing dogs from wolves

The existence of an animal-fact network can also explain how we can learn to classify things which seem superficially very similar into different categories – for example, why we say that wolfhounds are dogs rather than wolves, even though they resemble wolves more than they resemble, say, poodles.

In the previous chapter we saw that the process of learning to distinguish dogs from cats, or one Chinese person from another, is based on learning to attend to the differences in their appearance which we may not have noticed at first. Dogs look different from cats, and even different types of dogs look more like other types of dogs than they look like cats. Thus, as the cat-image pattern becomes more differentiated from the dog-image pattern, its connections with the pattern for the word "cat" in the animal-name network are strengthened, while its connections with the pattern for the word "dog" are weakened. It is therefore able to evoke the pattern for "cat" rather than "dog" whenever it occurs.

But we also learn to distinguish dogs from wolves, and this has to be done by a different process, since the distinction between them is not based solely on visual characteristics. If we classified animals purely on the basis of our visual images of them, then we would call a wolfhound a wolf rather than a dog, because the pattern for the visual image of the wolfhound is more similar to the pattern for wolves than to the pattern for dogs. How then can connectionist theory explain the fact that we say that wolfhounds are dogs rather than wolves?

The explanation makes use of the assumption that the image, name and fact networks for any given class of object are all interconnected. Thus when we encounter a specific animal – in our case, a wolfhound – the pattern activated on the animal-image network gives rise to a particular pattern on the animal-fact network as well as to one on the animal-name network. The units in the animal-fact patterns will then send their signals to the units of the animal-name network, arriving very shortly after the signals from the animal-image network that were sent there directly.

Now since the animal-fact network stores all the facts we know about animals, it includes such facts as "Poodles make good pets" or "Wolfhounds are useful for rounding up sheep," as well as sentences we have heard about individual animals, such as "Carol has a poodle that keeps following her around the house" or "Jack has a wolfhound that keeps following him around the yard." It also represents folkloric sentences such as "Who's afraid of the big, bad wolf?" or "They got lost in the forest and were eaten up by wolves." These facts tend to come in clusters: facts about dogs that stress their usefulness and friendliness for human beings, and facts about wolves that involve their fearsomeness and danger for people. Thus we learn to classify animals that look like dogs as either dogs or wolves according to the facts we know about them rather than merely their appearance.

Therefore, if we should encounter a wolfhound, both competing sets of connections with the animal-name network – the one from the animal-image network and the one from the animal-sentence network – are activated. The question now is which of these sets of connections is stronger. If the set of connections between the pattern for the wolfhound's image and the pattern for the word "wolf" is stronger, then we will call the wolfhound a wolf. If, on the other hand, it is the set of connections between the patterns for the facts about the wolfhound and the pattern for the word "dog" that is stronger, then the animal will be called a dog. But we generally call a wolfhound a dog, so the connections with the animal-fact network must be stronger. How does this come about?

Well, there are some differences between the way wolves look and the way wolfhounds look; otherwise we would not be able to distinguish them by appearance at all. Thus the question is how the small difference in appearance between wolfhounds and wolves can outweigh the much larger difference in appearance between wolfhounds and poodles. It is here that the connections *among* the various networks display their importance. Not only are both the animal-image and the animal-fact networks connected to the animal-name network, but the two networks are connected to each other as well – the sight of an animal can evoke the facts we know about it, and facts about an animal can evoke its image. Moreover, the connections between the various networks can change the structure of the patterns within the networks just as input from the outside can.

How might this work? We have just seen that the pattern for the wolfhound image is connected to a very different set of animal-fact patterns than the pattern for the wolf image, while the patterns for the wolfhound and the poodle images are connected to much more similar sets of fact patterns. We saw in the previous chapter how learning different names for dogs and cats makes the patterns for their images more different from each other as well. Now the same thing can happen through learning different facts. When we learn facts that show how different wolfhounds are from wolves, the small differences between the way they look become more important. Since we may want to pet a wolfhound but not a wolf, we need to be able to distinguish them by appearance, so we pay more attention to the differences in the way they look, and thus the patterns for the wolf and the wolfhound images also become more differentiated. As a result, the wolfhound-image pattern only weakly evokes the pattern for the word "wolf," while the wolfhound-fact patterns strongly evoke the word "dog." Thus the pattern of connections *among* the networks allows the encounter with the wolfhound to evoke the word "dog" more strongly than the word "wolf," and we call the wolfhound a dog.

Linguistic and emotional associations

The existence of the animal-fact network can also help explain our immediate response of "dog" when we hear "cat" on a word-association test. This association cannot be explained by semantic network theory at all. There are many different animals that are all linked to the animal node in a semantic network, so these links cannot explain why we say "dog" rather than "horse" or "bear" or "lion" when we are asked to say the first word we think of when we hear "cat." Indeed, since cats are more closely related to lions than to dogs, a link-based semantic network theory should predict that people would say "lion" or "tiger" rather than "dog" when they hear "cat." But we don't, so there must be some other way of explaining this response.

Here too we can use the animal-fact network to explain our response of "dog" rather than "lion" when we hear "cat." Although we know the fact that cats are more closely related biologically to lions than to dogs, we also know many facts about cats and dogs that point up the similarities between them. Both of them can be house pets, and both are friendly and often useful to people rather than feared wild animals such as lions and tigers. Here we can see a mechanism at work that is very similar to the one we used to group wolfhounds with poodles rather than wolves. Just as the facts we know about wolfhounds and poodles lead us to classify them as dogs, so the facts we know about dogs and cats lead us to classify them as tame rather than wild animals. Even though we also know the facts about the biological relatedness of cats to tigers and dogs to wolves, the facts about our own human concerns with tame animals as opposed to wild ones are much stronger in our animal-fact network because they are more important to us. Therefore the patterns for these human-related facts are more easily activated, and this is one reason why we say "dog" rather than "lion" in response to "cat."

But it seems to me that there may be another factor at work here, based on yet another network in our mind – what we might call a catchphrase network. The words "cat" and "dog" are found together in some very common phrases – cliches, if you will – such as "raining cats and dogs" or "fighting like cats and dogs." These phrases are fleetingly activated when you hear "cat," and they in turn activate "dog." None of these phrases is activated strongly enough to reach the level of awareness, but each of them activates the word "dog," and this is another reason for you to say "dog."

Here again we see how our system of interconnected networks can support many different kinds of associations between concepts. The word "cat" activates the phrase "cats and dogs" in the catchphrase network, the sentence "Cats are animals" in the animal-fact network, the sound "meow" in the animal-sound network, the image of a cat in the animal-image network, and, for those of us who have a pet cat, a great deal of specific information about our own particular pet.

As we have seen, a response appropriate to the question we have just been asked or the remark we have just heard is generally most strongly activated. Thus we tend to respond in accordance with the social demands of the situation we are in. This is what makes most human communication a cooperative enterprise. We may not give the answer that the questioner hoped for – we may say, "No, I can't do what you want" rather than "Yes, I'll do it" – but we usually answer the question that has been asked rather than some other question.

There are, however, situations where we are unable to be so cooperative, such as great emotional distress. In such cases the information that is most strongly activated is likely to be related to what is causing us distress rather than what the other person wants to know. We may therefore respond in a way that does not seem relevant to the present conversation.

If you are worried about your lost cat Stripey, for example, you are likely to say "Stripey" or "lost" rather than "dog" when the word-association tester says "cat." This is, in fact, one of the ways that the word-association test is used to detect areas that may be causing emotional distress.

Representing family relationships

Now let's see how our networks can be used to represent the relationships between family members without any links labeled "mother of", "father of", "son of" or "daughter of." One of these networks is a people-name network that contains the names of the people we know and the people we know about. There are probably separate networks for our acquaintances and the people we know about from history, but they should work more or less the same way. It is simplest to use an historical example for our discussion, since this information is shared by most people in the same culture. For people familiar with the Bible, the people-name network includes the names of the Old Testament characters Jacob, Leah, Simon and Dinah. Since we are claiming that there are no labels on the links between the names, how do we know in what way they are related to one another?

Well, just as there's an animal-fact network, there's also a family-fact network. It contains facts like "Jacob is the father of Simon" and "Leah is the mother of Dinah," as well as "Simon is the son of Leah" and "Dinah is the daughter of Jacob." Let's say I'm thinking of Jacob and trying to remember if he had a daughter in addition to his twelve sons. Then the name "Jacob" in my people-name network activates the fact "Dinah is the daughter of Jacob" in the family-fact network, and I recall that Jacob did indeed have a daughter.

But let's say that now I want to know what sort of relationship exists between Simon and Dinah. If I recall the verse in which Simon refers to Dinah as his sister, then I already have this fact stored in my historical family-fact network. But even if I do not recall this verse, I can use the information I do have about them and make a deduction from it according to some rules. How can this sort of deduction take place in neural networks?

We've already seen how the process of deduction is explained by semantic network theory. That theory suggests a way of retrieving the information that robins can fly without a direct link between "robin" and "can fly" in the network. Since there is supposed to be a link between "robin" and "bird" and another one between "bird" and "can fly," moving along these two links brings us the information that robins can fly. It was even shown by experiments that the process of judging whether a sentence such as "Robins can fly" is true or false takes longer than making the same judgment about a sentence such as "Birds can fly." This result was interpreted by the semantic network theorists as support for their claim that there are more links in the chain connecting the nodes in the first case than in the second. How can we explain this experimental result if we claim that these facts are stored in connectionist networks rather than semantic tree networks?

What we can say is that there are connections between the sentences in the various fact networks, just as there are connections between the words in the various word networks. For example, the two sentences "Robins are birds" and "Birds can fly," when activated together, lead directly to the sentence "Robins can fly." This is true whether or not the sentence "Robins can fly" is already represented in one of our fact networks. If it is not there yet, then a schema is activated that can be represented by the rule "If As are Bs and Bs do X, then As do X." This rule is not necessarily explicitly represented as a sentence in our mind – it is just a way of putting into words some mental process that takes a sentence of the form "As are Bs" and a sentence of the form "Bs do X" and produces the new sentence "As do X." This is an automatic process that does not require any explicit knowledge of the rule. Logicians have formulated some processes of this sort into explicit rules, but this does not mean that the rules are present as explicit sentences in the minds of people who are not logicians.

Now let's see how this explanation can help us understand how we figure out the relationship between Simon and Dinah. Well, the family-fact network works similarly to the animal-fact network, but it also contains some more complicated rules that may actually be present in the form of a tree – a "family tree" – in another network that might be called a family-tree network. When I consider the relationship between Simon and Dinah, I activate my family-tree network, where I have a schematic model of a family tree with Jacob and Leah as the father and mother, and Simon and Dinah as their son and daughter. I already have the general knowledge that if two people are the son and daughter of the same mother and father, then they are brother and sister, so the family-tree model of these four people automatically sprouts a brother–sister link between Simon and Dinah, and this creates a pattern of activity in my family-fact network representing the new fact, "Dinah is the sister of

Simon." This fact is then added to the family-fact network, and the next time I think about it I won't have to go through this whole process again, because I already know it.

It is important here to avoid confusing the semantic tree networks we are representing with the connectionist networks we use to represent them. The family trees we are representing actually resemble networks, so it is easier to confuse them with the neural networks that represent them than in cases where the networks represent other things, such as dogs or the word "dog."

To keep these two levels distinct, we must remember that the tree models in the present example are the thing to be explained, not the explanation. Just as neural networks explain how we remember words and use them to talk about things, they also explain how we form models of relationships that we picture in the shape of family trees. The trees are models of the relationships between people, and so they are one way we can use to remember these relationships. This way is generally much more efficient than trying to remember a list of facts about these people, so we use them in our minds to help us remember. But the tree models are not physical tree networks in the brain any more than words are physically written in the brain. Both tree models and words exist in the brain only in the form of the neural networks I have been describing throughout this book.

8

Evidence for connectionist models

What evidence is there that our minds actually work in the way I have been describing? What sort of experiments have been done to support these new notions? Are there any new discoveries in biology supporting the connectionist claims?

First I will describe a sample of the many experiments performed to test connectionist models, and then I will present a fascinating new finding in biology, made independently of connectionist theory, that also supports this theory.

Experimental evidence

Although scientists have been working on developing connectionist models only since the 1970s, most of them test their models experimentally as soon as they begin to develop them, so that there is actually a great deal of experimental evidence for these models by now. In fact, there is a reciprocal relationship between the models and the experiments – the models provide ideas for experiments, while the results of the experiments often make it necessary to alter the models somewhat.

Connectionist models are tested by two different types of experiments. First it is necessary to see whether the models work at all. This is done by setting up a computer simulation of the particular human performance described by the model and seeing if the simulation produces something close to the results obtained in humans. For example, a test of a model of how people see things in three dimensions might be to program a computer to analyze a scene the way we think people do it. Then we give the computer some scenes to analyze and we look at the results of its analysis

to see whether it comes up with the same objects that a human being would find in the scene.

Analyzing two-dimensional scenes

Such a test might run as follows. We put a few simple geometrically shaped objects on a table, such as a cube, a cone, a pyramid and a cylinder, with some of them in front of some of the others, in such a way that each of the objects in the back is only partially visible. We then sketch or photograph this scene and present it to a computer equipped with vision sensors and rules for deducing the presence of three-dimensional objects from two-dimensional photographs or drawings. The test is to see whether the computer program comes up with the same list of the objects on the table that a human would.

Let's say the results of such a simulation are positive – that is, the simulation produces the same "behavior" we have observed in humans. In our case, this would mean that the computer listed the cube, the cone, the pyramid and the cylinder as being present in the scene. Specifically, it would mean that the program could "see" the cone as a cone even though the cube was in front of it and the cone was only partially visible.

It is important to understand what the program has to do in order to act "human." One of the things it has to do is jump to conclusions which are not based on pure logic, just as humans tend to do. In our particular example, all we or the computer can really see is an object that looks like part of a cone with a cube in front of it. We have no way of knowing for sure that the part of the object that is hidden behind the cube is actually shaped so as to form a complete cone. A slice may well be cut off one side of the cone where it's hidden behind the cube, yet we unhesitatingly "see" a whole, complete cone and are prepared to testify that there was a cone on the table. In order to act "human," then, the computer too must draw the unsubstantiated conclusion that there is actually a complete cone on the table. In other words, it must be programmed to come up with the same conclusions that people do, even when these are not strictly logical.

Artificial intelligence and "natural" simulations

But even if the computer is successful at this task, it does not necessarily mean that the human mind actually does the task the same way that we have set up the computer simulation to do it. Perhaps people do the task by a different process than the simulation, yet still get the same result. Scientists in the field of "artificial intelligence," or AI, who are interested only in getting computers to do the sort of thing that humans do, would not care about this. All they want is to model human *actions*. But cognitive scientists, who are interested in the way the human mind works, demand more. What we want to know is whether the computer simulation might actually model the *process* by which human beings see three-dimensional objects or learn to pronounce words.

There are several means we could use to find out whether this could be the case. One of these is analyzing errors. If we could show that the simulation makes the same learning errors as a child learning a new task or an adult performing a well-known one, then we would have good reason to believe that the simulation is compatible with what actually goes on in our mind and our brain when we learn or perform this task.

Another often-used method is analyzing reaction times – that is, the amount of time it takes people to perform various tasks. Let's say one theory predicts that recognizing a sphere should take longer than recognizing a cube, while a second theory predicts that recognizing a cube should take longer. Then an experiment showing that it takes more time for people to recognize a sphere than a cube would be evidence in favor of the first theory and against the second one.

A computer learns to talk

One of the most intriguing computer models of this sort, which was designed to show how a connectionist network can simulate the way babies learn to talk, is a program called NETTALK created by Terry Sejnowski and Charlie Rosenberg. This program does not simulate the process of learning to attach names to objects, as described in Chapter 6, but rather the process of learning to pronounce words so that they sound like the words heard in the environment. For the baby, the environment is the family; for the computer model, it is the words fed into the computer by the programmer.

NETTALK begins with a three-layer network of units. The input layer gets its input from a series of words read aloud to it. The output layer produces sounds. These two layers are connected by an intermediate layer of units to provide the capacity for change. At first the connections between the input layer and the intermediate layer, as well as those between the intermediate layer and the output layer, are all random. When the input layer hears "cat," for example, it sends completely random information to the output layer, so that the sound it produces may well be "vrrip."

In order for NETTALK to learn the proper pronunciation of "cat," it is provided with a mechanism for correcting itself. It "hears" its own output and compares it with the input it has received. It then modifies the connections between the units in the different layers so as to slightly narrow the gap between the input and the output. This is a very slow process that takes many rounds of hearing input words, producing output words, and repeatedly modifying the connections so as to make the output closer to the input.

In our example, after hearing "cat" and comparing this sound with its own output of "vrrip," NETTALK may modify its connections so as to produce "vrrap" instead on the second round. The third round might be "vrap," the fourth "rap," and so on. Finally – actually after many more rounds than in this simplified example – NETTALK will end up producing "cat" as output.

It is intriguing to listen to a speaker attached to the computer modeling NETTALK and actually hear the output of the program transformed into real sounds. At first the output sounds like gobbledygook, totally unrelated to any real words, yet it ends up producing a string of clearly recognizable words. The initial computer output does not sound much like a baby producing word-like sounds before it has learned to say real words, but there is a real parallel here. The word-like sounds produced by babies begin to resemble the sounds of the language they hear around them long before the babies learn any actual words. Moreover, the words babies produce often distort many of the sounds of the words they are clearly trying to imitate, and it generally takes them quite a long time to produce words that sound "correct."

NETTALK is thus a good example of a model that passes the test of "learning" successfully and in a way that somewhat resembles the way human beings learn. Another such model is one that learns the past tenses of verbs similarly to the way children do. This model was devised by David Rumelhart and James McClelland, with the intention of simulating the pattern of correct usage and errors in children's speech as they learn to use the past tense of verbs.

A computer learns the past tense

Psychologists observing children's speech have noted that at first they use the past-tense forms they hear in their environment, using "liked," for example, as the past tense of "like," and "fell" as the past tense of "fall." At this stage most of the few verbs children are able to produce are irregular ones, such "do–did" and "make–made," so they learn each pair separately.

Later on, however, children have more experience with regular verbs, such as "watch–watched", "kick–kicked", "play–played", "try–tried", and "fix–fixed." At this point some children "regularize" the irregular forms, saying "falled" or "felled" as the past tense of "fall," and "bringed" or "broughted" as the past tense of "bring," even though they have already produced the correct past-tense forms "fell" and "brought." At yet a later stage they use the conventional forms once again, going back to "fell" and "brought."

How was this pattern of correct usage and error explained before the connectionist model was devised? The prevalent explanation has been the linguistic rule-following theory. According to this theory, children at first simply imitate what they hear, but after a while they learn the rule "Put '-ed' after the present tense of the verb to form the past tense." Linguists of the rule-following school, the most well known of whom is Noam Chomsky, do not claim that this rule is formulated by the children as an explicit sentence, but they do insist that it is represented in their mind in some way.

Rumelhart and McClelland's connectionist model was thus designed to show how people could seem to be following a rule – in this case, adding "-ed" to form the past tense of verbs – without actually doing so. In the computer simulation the input to the system consists of pairs of the present- and past-tense forms of everyday verbs, including regular forms such as "like–liked" and irregular forms such as "fall–fell." In the first rounds the verbs in the input are the most frequently used ones, and then less frequently used verbs are added, to model the verbs children are most likely to hear.

The first rounds of the computer simulation thus include mostly irregular verbs, such as "come–came", "give–gave", "tell–told", "eat–ate", "break–broke," and "go–went." Since at this point there are few regularities in the formation of the past tense from the present tense, the computer simulation produces only slightly overlapping patterns of activity for the various present and past tense pairs on the list. Thus the model simulates the way children learn each of these pairs separately at first.

In the second stage the computer is presented with the verbs children learn somewhat later. Since this group contains a large proportion of regular verbs ending in "-ed," the past tenses are more similar to each other than they were in the first stage, and so the simulation stores them as patterns of activity that overlap to a greater extent than those representing the irregular verbs. The parts of the patterns that represent the "-ed" endings of the regular verbs thus become strengthened by being repeated in many different verbs. The irregular past tenses, in contrast, occur in only one or two verbs each. Even though many of these irregular verbs occur more frequently than any one individual regular verb, there are more past tenses formed by adding "-ed" than any other kind of past tense.

Thus, at this stage, the dominant response is adding "-ed," and this tendency is stronger than the tendency to produce the irregular past-tense forms learned earlier. The process of putting a verb in the past tense activates the whole past-tense network, and the strength of the overlapping "-ed" endings makes the "-ed" pattern an attractor (as described in Chapter 5), so that whenever the network is activated, the "-ed" ending is attached to the verb. Sometimes the activation of the "-ed" ending occurs before the correct past-tense form of the irregular verb is activated, and then the "-ed" is added to the present-tense form directly, leading to "errors" like "bringed" and "goed." At other times the already-learned past-tense form of these verbs happens to be more strongly activated, and then the subsequent activation of the attractor leads the "-ed" ending to be added to this past-tense form, producing "errors" like "broughted" and "wented."

We can thus see that the connectionist model provides a better explanation of children's behavior at this stage than the rule-following theory. If children were really following a rule, then they should always add the "-ed" to the present-tense form of irregular verbs or else they should always add it to the past-tense form. But the evidence from children's actual speech shows that the very same child may say "broughted" in one sentence and "bringed" in the next. This is hard to explain on the basis of a rule-following theory. It is more easily explained by a connectionist model in which the activation of different patterns depends on a combination of the strengths of various connections, based on long-term learning, together with the temporary state of activation, based on the input just received.

Here is an example of the way this might work in the case of "bringed" and "broughted." Let's say Susie is in the second stage and has learned many regular verbs. When she wants to tell her mother that her father has given her a new box of crayons, she says, "Mommy, Daddy bringed me some crayons." But her mother, who is trying to provide Susie with a model of accepted speech without directly correcting her, says, "Oh, how nice, you must be happy that Daddy brought you the crayons." This activates the past-tense form "brought" in Susie's past-tense network, so now, when she continues her conversation with her mother, she says, "He broughted me a yellow one and a red one and a blue one." The long-term tendency to add "-ed" based on the strength of the overlapping "-ed" endings remains, but the recent activation of "brought" causes "broughted" to win out over "bringed."

But then how do children learn to use the conventional past-tense forms of verbs in the third stage? According to the rule-following theory, they learn that, alongside the rules, there are also exceptions to the rules, and they then learn each of the exceptions separately. Now let us see how the connectionist model explains the third stage of past-tense formation.

In the third set of rounds, the input to the computer model consists of a wide variety of verbs, both regular and irregular. At this stage most of the verbs are regular, since this is the case for the English language as a whole. Now, as we saw in Chapter 6, learning takes place through continual, gradual changes in the strengths of the connections between the neurons. In the first stage, when the few verbs learned could be represented by only slightly overlapping patterns of activity, there was very little interference among the patterns – each past tense was represented separately and therefore reproduced correctly. In the second stage, when more verbs were learned, the connections that were strengthened most in the past-tense network were the ones found in the majority of regular verbs. At this point there was a great deal of overlap, and since the changes in the connections take place very slowly, the strongly overlapping regularities overwhelmed the fine distinctions between the different irregular verbs.

Now, in the third stage, as the child continues to hear both regular and irregular verbs over and over again, the individual connections between the individual present and past tenses of the irregular verbs are gradually strengthened as well. Eventually these connections become strong enough so that the child can produce the conventional past tense for each irregular verb. At the same time, the child will continue to add the regular "-ed" ending to any verb for which a special past tense has not been learned, since adding this ending remains the strongest general tendency.

The Rumelhart and McClelland model thus shows how behavior which seems to be the product of following a rule can actually result from gradual changes in the connections of a network, which are strengthened most by regular input, thus producing regular output as well. This does not mean that people do not ever learn explicit rules, of course; it only means that they often act in ways that seem to show that they are following rules even though they do not actually know the rules.

A computer prefers words

Not all computer models of mental processes were developed to show how familiar human activities, such as learning to talk, might take place. Some of them were intended to explain interesting effects that no one would ever have noticed if laboratory experiments had not been designed to elicit them. In fact, one of the earliest connectionist models, also developed by McClelland and Rumelhart, was designed to test an intriguing experimental result known as the "word-superiority effect."

The word-superiority effect was discovered experimentally in the early days of cognitive psychology, when the most widely used methods for finding out about our mental processes were measuring reaction time – for example, how long it takes to say whether "bore" or "bork" is a word – and error rates – for example, whether you are more likely to make a mistake reading "bork" than "bore." In those early days cognitive psychologists were still spending a great deal of time trying to prove that the mind has an easier time dealing with meaningful things than with meaningless things, as they had not yet succeeded in overthrowing the behaviorist tenet that meaning doesn't matter. Now words are things with meaning, while letters are things without meaning. If the mind doesn't care about meaning, then we should be able to perceive letters faster than words, because each word is made up of several letters. If the mind does care about meaning, then we might actually be able to perceive smallish words better than individual letters.

The experiments designed to test this used the error-rate paradigm. Participants were presented with either an isolated letter, or a four-letter word (an ordinary word, not an expletive) containing that letter, or a four-letter nonword (a string of four letters that does not constitute a word) also containing the letter. For example, let's say one of the words was "cork." Then some participants saw the letter "k," some saw the word "cork" and some saw the nonword "bork." These letters or letter strings were presented for a very short time, so that it would be hard to see them and the reader would make a lot of errors, thus giving the experimenters a substantial error rate to measure in order to compare performance for the letters, words and nonwords. The task for the participants was simply to say which one of two given letters was in the presentation they had just seen.

In the above example the two letters were "k" and "d," so participants would have to say if they had just seen a "k" or a "d." The "d" was chosen as the alternative because it also makes a word if substituted for the "k" in "cork," producing the word "cord," and a nonword when substituted for the "k" in "bork," producing the nonword "bord." It would not be a good idea to have alternative letters where one would produce a word, such as "cork," while the other would produce a nonword, such as "corb," because then the reader could guess that there was a "k" and not a "b" just by remembering that there had been a word rather than a nonword on the screen.

The experiment produced the results predicted by those who claimed that the mind cares about meaning. Not only were people more accurate at saying that they had seen a "k" and not a "d" when they saw "cork" than when they saw "bork," they were also more accurate in their choice when they saw "cork" than when they saw the "k" alone. This human ability to perceive words better than isolated letters was dubbed the "word-superiority effect."

When McClelland and Rumelhart designed their first connectionist computer model, it was this effect that they intended it to explain. By this time, in the late 1970s, a number of studies had been done expanding the work just described. These studies showed a hierarchy in the perception of both words and nonwords: While words are more easily perceived than nonwords, not all words are equally easily perceived, and not all nonwords pose the same degree of difficulty. Among words, frequency is an important factor: More frequent words are perceived more easily than less frequent ones. Thus "bald," for example, would be perceived more easily than "balk."

Among nonwords, resemblance to real words seems to be the important factor. A nonword like "bork" is known as a legal nonword because it could be a word in English if some group decided to make use of it, just as teenage slang has given a meaning to the previously meaningless string "dork." Such legal nonwords are more easily perceived than illegal nonwords like "sbek," which could never be an English word. But even a nonword like "sbek" could conceivably be a word in some other human language, because it is pronounceable, in contrast to unpronounceable strings like "ksvp." And indeed, pronounceable nonwords are more easily perceived than unpronounceable ones.

McClelland and Rumelhart therefore designed their model to explain this entire hierarchy of effects: frequent words > nonfrequent words > legal nonwords > illegal but pronounceable nonwords > unpronounceable nonwords.

The model they proposed is a simple connectionist one in which every letter has two activating connections with every word containing that letter – one in each direction. For example, seeing the letter "c" activates the word "cork," while thinking of the word "cork" activates the letter "c." Words that are seen more frequently develop stronger connections with the letters they contain than less frequent words, according to the process explained in Chapter 6. Thus the letter "c" will activate "coat" more than it will activate "cork." This explains why frequent words are read more easily than nonfrequent words.

But why should legal nonwords such as "bork" be more easily read than illegal but still pronounceable nonwords such as "srop"? McClelland and Rumelhart's model explains this finding quite ingeniously, by postulating that fragments of words are also represented in a network. Now the legal nonword "bork" contains two three-letter fragments, "bor" and "ork," both of which are found in actual four-letter English words – "bor" in "bore" and "born," for example, and "ork" in "cork" and "fork." Seeing "bork" therefore partially activates these words in our word network, and they in turn provide more activation for the letters in "bork," thus making it easier to read. In the case of "srop," in contrast, while "rop" is part of "drop" and "crop," "sro" is not part of any four-letter word in English. Thus the letters in "srop" get some extra activation from only one of its two three-letter fragments, and it is therefore harder to read than "bork."

The explanation of the finding that pronounceable nonwords are easier to read than unpronounceable ones is very similar. Now we are comparing four-letter strings like "srop" to strings like "srkp." The string "srop" at least gets some activation from "rop," but "srkp" gets no activation from either of its three-letter fragments, as neither "srk" nor "rkp" is part of any English word.

McClelland and Rumelhart then implemented their model in a computer program and tested the program with experiments similar to those that had provided the findings for the relative ease of reading various words or letter strings and perceiving the individual letters in the words or strings. When these same experiments were peformed on human participants as well, the results for the computer program and the humans were quite similar, providing excellent evidence that this model may capture the way we actually perform these tasks.

The models I have described in this chapter thus do both of the things required of a model of how the mind works: They produce the behavior seen in human beings, and in the process they produce some of the errors seen in human behavior as well, making it plausible that the mechanism they are modeling is similar to the one used by the human brain.

Evidence from the physiology of smell

Another type of evidence that can support the connectionist theory comes from physiological discoveries about the way neurons work together to produce sensations. Physiologists have been studying the neuronal basis of the various human senses for several decades. They have discovered that we see, hear, smell, taste and feel things through neurons on the surface of our various sense organs, called receptors. These receptors take the electromagnetic vibrations of light, the vibrations in the air that we call sound, and the shapes of the molecules that cause odor, and translate them into the electrical impulses used as signals by the neurons (this signaling process was described in Chapter 3).

After decades of progress in understanding how we see and hear, physiologists have recently made considerable advances in analyzing the sense of smell. They have known for a long time that there are receptors in the nose that are sensitive to odor molecules, but they did not understand how people can discriminate more than ten thousand distinct odors even though there are only about a thousand odor receptors. But in 1998 and 1999 two research groups, one led by Linda Buck and the other by Randall Reed, discovered that each odor molecule, called an "odorant," binds to a different array of receptors in the nose, with each receptor being part of several different arrays.

For example, the odorant octanol, which produces a sweet, orange-rose scent, binds to the receptors labeled S1, S18, S19, S41, S46, S51, S79, and S83, while the odorant octanoic acid, which produces a sour, sweaty smell, binds to four of the same receptors – S18, S19, S41, and S51 – but not the other four. The difference between the array of eight receptors and the array of four out of these eight is sufficient to cause this totally different sensation of smell.

Moreover, an array similar to that for the sour odor of octanoic acid, containing the same receptors except for S46 and S83, is stimulated by the odorant heptanoic acid, which produces a very similar smell. The odorant heptanol, which stimulates only one receptor – S19 – in common with the other three, but adds two new receptors – S3 and S25 – produces a very different odor – herbal, woody and reminiscent of violets. And the odorant hexanol, which stimulates just these last two receptors, produces a similar odor, also herbal and woody, but reminiscent of cognac and whiskey rather than violets.

These arrays of sensory receptors, with similar arrays producing similar odors and more different arrays leading to more different odors, seem very much like the patterns of activity representing different concepts in connectionist theory. The odor receptor arrays thus provide hard scientific evidence supporting the theory that there are similar types of patterns within the brain, which could underlie the concepts in our minds.

Two different types of memory

Until now we have been discussing how we learn and organize our general knowledge – the names and properties of things, such as dogs and cats, or mothers and fathers, and the relations between them. All these may be called permanent memories, as we generally retain this sort of knowledge throughout life, and it changes only occasionally, such as when we hear about a new type of mother known as a "surrogate mother." But when we think of all this knowledge as a type of memory, we immediately begin to think of the other kinds of things we need to remember – namely, the specific things that happen to us or that we do in our daily life, and the things we are intending to do. These may be called temporary memories, as they involve specific occasions and do not need to be remembered for a long time. In this chapter I discuss some of the essential differences between the two types of memory and describe what is known about the way they are embodied in different types of neural networks in the brain.

Differences between the two types of memory

Permanent and temporary memories differ in many ways. For example, the fact that a kitchen is a place to eat is part of my permanent memory, while the fact that I opened up a new box of cornflakes this morning is part of my temporary memory. Permanent memory is time- and person-independent: The truth of the statement "A kitchen is a place to eat" does not change if I say it today or tomorrow, and it does not depend on whether I say it or you say it. In contrast, if the statement "I opened up the last box of cornflakes today" is true on Monday, it is rather unlikely to be true on Tuesday as well. Moreover, the fact that this statement is true

"Oh no, not again!"

when I say it does not have any bearing on whether or not it is true when you say it.

The type of connectionist model I have been describing so far cannot explain how we remember things that are only relevant temporarily, such as whether I took my medicine this morning or whether you deposited the check in the bank. The reason for this is that learning takes a long time in connectionist networks. As described in Chapter 6, the strengths of the connections between the various concepts change very slowly and gradually, so that new information can be added without disrupting old information. But this can't be true of things we remember immediately, such as my opening the last box of cornflakes. This is something I have to remember long enough to buy a new box, but not any longer than that, so that I don't keep on buying new boxes. Thus, in addition to being able to remember things quickly, the mechanism underlying temporary memory must enable us to forget things rapidly as well.

Opening the cornflakes box may not be a very important thing to remember, but there are many events that are quite important for us to be able to remember immediately and forget not very much later. Consider, for example, the statement "I took my medicine this morning." On the one hand, I have to remember that this statement is true as soon as I have actually taken my medicine in order that I should not take it again on the same morning. On the other hand, if this statement was true on Monday, I must not continue to believe it in that form – that is, in the form "I took my medicine this morning" – on Tuesday, as then I would not take my medicine on Tuesday. What is needed here is an updating mechanism that works very quickly, and that distinguishes one point in time from another.

The reason the connectionist model cannot explain this sort of rapid learning and almost equally rapid forgetting is that it works by making very small changes to the strengths of the connections between the neurons. The changes must be small so as not to disrupt the information already stored in the network, but this means that many repetitions of the new information are required in order for it to be stored. This is very good for general information – after all, it makes sense to believe that something is true in general only if we encounter it a number of times. It is not so useful, however, for knowing what day it is today or remembering what we ate for breakfast. This type of memory requires another sort of mechanism altogether.

A connectionist model of temporary memory

A very interesting new model for how temporary memory works has recently been suggested by the same James McClelland who has worked on connectionist modeling of permanent memory. He and some colleagues, including Bruce McNaughton, proposed a model of temporary memory which actually works much more like the traditional way people have always thought of memory working, with different items of information stored locally in different places. Although this model of temporary memory also involves networks of connected units, it is different in several ways from the connectionist model of permanent memory. I will now describe the most important differences between the two models.

First of all, the two types of memory are stored in two different parts of the brain, with a different structure of connections between the units. In our discussion of general-knowledge networks we saw that there are many different networks for different types of knowledge, but they have the same basic structure, with all the units connected to all the other units either directly or at one or two removes. Moreover, all the pieces of information in a particular network are stored in a distributed way over the units of that network, and each network stores a different type of information – names, or shapes, or sounds, or odors.

Although the neurons making up the network for storing temporary memories are the same sort of neurons, and the connections between one

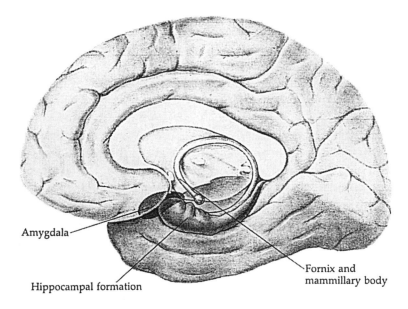

Amygdala

Hippocampal formation

Fornix and
mammillary body

Figure 9.1. The cortex, which stores permanent knowledge, surrounds the inner parts of the brain, including the hippocampus, which stores temporary memories and helps make some of them permanent.

neuron and the next work the same way, the structure of the network as a whole is very different. Permanent memories are stored in networks in the outside part of the brain, just under the skull, called the "cortex." The cortex is divided up into different areas on the basis of the way the information gets in from the senses, so we have image networks connected to the areas that process visual information, name networks connected to the areas that process language, and many others, as described in Chapter 5.

Temporary memories, in contrast, are stored in an area in the internal part of the brain (see Figure 9.1), consisting of the hippocampus and some related structures. Since the hippocampus is the most important of these structures, the area is generally referred to just as the "hippocampus."

Another difference between the two types of memory is that each general fact is distributed over all the neurons of a particular network, as described in Chapter 5, while temporary memories are stored locally. That is, each temporary memory takes up only a small number of neurons, in contrast to the large number taken up by the permanent memories in the connectionist networks. As a result, there is very little interference between the memories in the temporary network, as each one is in a different "place." Large, rapid changes can therefore be made in the strengths of the connections between the neurons in the cluster storing an individual memory, and this is what allows us to remember something that happened only once.

From temporary to permanent memory

A temporary memory will last only for a short period of time, however, unless something happens to embed it in a permanent store in one of the networks in the cortex. The reason memories in the hippocampus can be stored only for a short time is that, since each memory occupies a different cluster of units, only a limited number of memories can be stored before they start interfering with one another. When a similar temporary memory needs to be stored – for example, when I have to remember "I took my medicine this morning" for Tuesday instead of Monday – today's memory tends to wipe out yesterday's memory. This is because it is typically stored over the same cluster of neurons, and the changes between the connections have to be quick and large, thus effectively erasing the older memory.

How then do I remember that I generally have cereal for breakfast, even though I can no longer remember whether I had oatmeal or cornflakes on Wednesday two weeks ago? This general memory depends on the fact that there are many connections between the temporary and the permanent networks. Whenever an episode occurs and is recorded in a temporary network, it activates the connections with the appropriate permanent networks – in this case, the visual, taste and smell networks that store memories about food. This causes some small changes in the connections between the neurons in these networks, as described in Chapter 6. The aspects of the episode that occur only once – such as the fact that I ate oatmeal rather than cornflakes today, or the fact that the phone rang while I was eating and the oatmeal got cold – cause such small changes in the permanent network that the episode cannot be remembered weeks later. However, those aspects that occur repeatedly – such as the fact that I eat my cereal from a yellow bowl, or the fact that I read the newspaper while I am eating breakfast – cause repeated gradual changes in the same permanent networks, thus enabling these memories to be stored for a long time.

Still, there are also some things that happened to you only once that you nevertheless remember for the rest of your life. How is this possible according to the two-part memory system I have been describing? One suggestion that has been proposed by cognitive scientists is that these are important episodes you keep on thinking about many times after they have occurred. Each time you think about your first date, for example, the connections between the temporary and the permanent networks cause gradual changes in the connections between the neurons in the latter, in exactly the same way that repeated occurrences of the same type of event do. Thus your repeated rehearsal of the event strengthens its representation in your permanent memory in the same way as repeated occurrences of an event do, and that is why only important events – the sort that you rehearse to yourself or describe to others over and over – are remembered for a long time.

This description suggests an explanation for the fact that we sometimes remember unimportant aspects of important events, such as what we were doing when we found out that John Kennedy had been shot (for those of us who are old enough to have been around at the time; younger readers may substitute their own clearly etched memories). An earlier explanation that had been offered is that these are "flashbulb" memories – that the event is so traumatic that we remember everything that was happening at the time. But this can't be true, because we don't remember, for instance, what we were wearing when we heard the news. What we remember is precisely what we kept telling everybody when we discussed the event – namely, what we were doing then.

Still, there are some interesting differences between what we remember because it recurs often and what we remember because we rehearse it. When the same sort of event occurs over and over again – when I eat cereal for breakfast almost every day – what is strengthened is the memory for the actual events, so my permanent memory generally ends up being fairly accurate. In the case of the one-time occurrence – such as your first date – what is strengthened is not the memory of the event itself but the memory of the way you described the event to yourself and others. Thus the memory stored in the permanent cortical networks may end up being different in many respects from what actually happened, as your rehearsals of the event may focus on, say, its more pleasant aspects, ignoring the unpleasant ones.

From permanent to temporary memory

Interestingly, the connections between the two types of memory are not solely one-way – not only do temporary memories accumulate to become permanent ones, but permanent memories can also affect temporary ones. How could this happen? How can old memories change the way I remember something that just occurred recently?

The reason that my memory of something that happened recently can be affected by past memories is that the past memories are so much more strongly embedded in their networks. Let us say, for example, that I usually pour my cereal into the bowl first and then add the milk. On Tuesday, however, for some reason, I poured the milk into my bowl before pouring in the cereal. How will I remember this on Wednesday? At that time I may well believe that on the previous day I poured in the cereal first, as I usually do. But why should I?

My temporary memory of pouring the milk into the cereal bowl first on Tuesday may be eclipsed by my memory of generally pouring in the cereal first because my permanent memory is based on many incidents in which I did just that. At the same time that my temporary memory network is storing the Tuesday incident, this incident is also activating the permanent network in which things that I do every day are stored. Since the changes here occur very gradually, any changes due to my unusual behavior on Tuesday will be very slight, and so they will not appreciably change my memory for what I generally do. But then when I try on Wednesday to remember what happened on Tuesday, the fading trace of that isolated incident in my temporary memory may well be over-ridden by the much stronger connections in my permanent memory of what I usually do.

This explains why we ordinarily remember individual events as being more like the typical event of that kind than they actually were. Dramatically unusual occurrences – say, the telephone call that interrupted my breakfast being the one that I had been expecting for weeks – are an exception. In such cases the unusual event will be remembered through the rehearsal process described above.

We have thus seen how the special nature of our temporary memory store makes it possible for us to both remember quickly and forget quickly those things that we have no need to remember for a long time, or that we need to forget in order to prevent them from interfering with similar but updated information that we need to remember now. The differences between the two types of memory processes are firmly rooted in differences between the areas of the brain where they take place, while the similarities between them are grounded in the similarity of the connections between individual neurons in all parts of the brain. This offers yet more evidence that using the structure of the brain as a source of theories and models about the workings of the mind can provide us with a deep understanding of our mental processes.

10

Coping with disaster

What happens when our networks are damaged? How do they cope with the impact of soft brain tissue against the hard skull in traffic accidents, or the strokes or neural degeneration that may occur in our later years?

A great deal of research has been done on the effect of strokes on our various mental abilities, and computer models have been designed to mimic this sort of damage, so I will have the most to say about how our networks cope with strokes. Traffic accidents tend to cause a very different sort of damage, to an area which has been difficult for connectionist theory to explore, and I will try to explain the problems involved. I will have only a few words to say about the connectionist explanation of what happens in Alzheimer's disease, as very little is known about it.

What do strokes do to our networks?

So far we have seen how our knowledge networks are built up slowly and gradually over the years to provide us with ways of acting in the world, methods of distinguishing one thing from another, memories of what happened a long time ago and information about what has just occurred. What happens when these networks are damaged? When someone has a stroke – when an artery in the brain is blocked by a blood clot and the surrounding area no longer receives blood to nourish its neurons, so that many of them die all at once? Or when someone has a brain hemorrhage – when an artery in the brain bursts and floods the surrounding area with blood, again killing many neurons at once?

One very important point here is that it is the event of many neurons in a single area all dying at the same time that causes the damage. The death of a few neurons in any one area is actually quite commonplace and totally unnoticeable. Neurons in our brains die every day, and we are none the worse for it, even though not many new neurons grow after infancy. The reason our brains are so resilient, and can cope so easily with the daily loss of a few neurons here and there, is precisely the fact that the information in our networks is distributed over so many neurons. Since our memories and knowledge are not contained in individual neurons or small groups of them, but are distributed over whole networks, each individual connection between two neurons contains only a tiny part of the information, and the loss of a few such connections leaves the whole memory or piece of knowledge reasonably intact.

A stroke, however, whether caused by a clot or a hemorrhage, may end up killing quite a large number of neurons in the same network. In this case the damage will certainly be noticeable, but it will still be distributed. If the damage is extensive, one or more entire networks, such as the animal-name or the food-name network, may be totally destroyed, in which case the patient will not be able to say the name of any animals or any foods at all. If the damage is less severe, it will not totally destroy any particular piece of information in the network. The reason for this is that each piece of knowledge is stored over the whole network, and part of the network is still intact. What happens instead is that all the information in the network will be damaged to some extent, depending on the degree of damage to the network as a whole, rather than some bits being totally lost and some bits being totally retained.

For example, let's say Matilda's damaged network is the animal-name network we have become familiar with, and let's say that about a fifth of this network is affected. If the animal names were stored at random in individual neurons or groups of neurons, then we would expect Matilda to be unable to name about a fifth of the animal pictures she is shown on a test. If names of animals that are closely associated, such as cat and dog or cat and tiger, were stored together in a physically close set of neurons, we would expect Matilda to be unable to name all the animals in some particular group, say all the rodents or all the members of the cat family. But such outcomes of strokes have not been observed.

The usual case is actually quite different. Assuming that Matilda's case is typical, she will probably have some trouble naming all the animals she knows. When she is shown pictures of various animals, it will take her longer to name all of them than it would have taken her before. Moreover, she will probably have the least trouble naming familiar, well-known animals, such as cats or mice, while she may be totally unable to recall the names of unfamiliar animals, such as opossums or ocelots.

This occurs because a large number of the connections between the neurons in this network have been damaged, but a large number also remain intact. In the case of the very familiar names, the connections were all quite strong to begin with, so the ones that remain intact are sufficient to enable Matilda to recall the names. In the case of the unfamiliar names, the connections were never very strong, and so all of them working together were needed to allow the name to be activated when the picture is seen. Thus, when a large number of these connections are damaged, the remaining ones are not strong enough to enable Matilda to retrieve the names when she sees the pictures.

But what happens in intermediate cases, where the animals are moderately familiar, such as kangaroos or giraffes? We recall from Chapter 5 that each network is connected to many others, and that a particular pattern in one network can be activated by patterns in several other networks that are active at the same time. In that chapter we presented the example of an unimpaired person who is unable to recall the name of a big cat in a picture, and we showed how the first letter "o" might help the person recall the name "ocelot."

What happens to stroke patients when they try to recall moderately familiar words is often quite similar. In our example, Matilda may be unable to recall the word "donkey" when presented with the picture of a donkey. She may even say "horse," just like the child learning to speak that we described in Chapter 6, because only the well-used connections associated with "horse" are strong enough to activate a word. In such a case the first letter "d" may help Matilda recall the name "donkey."

Indeed, this particular type of cuing has been used for a long time by neuropsychologists examining the effects of brain damage on mental functions. A stroke patient who can recall "donkey" when presented with a picture of a donkey alone is considered in better shape than one who needs the first letter "d" as well as the picture, while a person who can recall "donkey" only with the aid of both the picture and the letter is still not as severely injured as one who cannot recall the word even with both cues. The theory that input from several networks provides more activation than input from only one thus helps us understand this well-known clinical finding.

Can stroke damage be repaired?

After the damage caused by a stroke has been assessed, the most important question for anyone facing this situation is whether it can be repaired. The answer to this question depends in part on the sheer extent of the damage, and in part on the particular networks and sets of connections that were damaged. The essential question is whether an alternate route – a "detour" – can be found that will permit the person to perform some important acts in cases where the usual route is now "out of order." If there are already connections in place that can serve as the basis for such a detour, even though they are weak because they were not used much before the stroke, they can often be strengthened and used in place of the damaged connections.

In our example of Matilda's stroke, in which the animal-name network itself was damaged, input from several other networks at the same time will probably be necessary to activate the names of animals, since individual neurons do not regenerate. However, Matilda can be encouraged to try to produce such additional input by herself, instead of waiting for someone else to provide her with cues. For example, she can be taught to try to think of as many facts as she can about the animal in the picture. Since we have been assuming for simplicity in this case that Matilda's animal-name network was the only one damaged in the stroke, she should be able to learn to do this. Then, when faced with the picture of a donkey, she will be able to activate sentences like "This animal eats hay" or "This animal carries loads" in her animal-fact network. The activation in this network could then work together with that in the animal-picture network to activate Matilda's animal-name network sufficiently to allow her to say "donkey."

In some cases the stroke may damage the connections between two networks rather than those within a particular network. Such cases may be treatable by a very similar method, even though the processes of activation between the networks are somewhat different. Let us say that the connection between the animal-picture network and the animal-name network has been damaged in Carlton's brain as the result of a stroke. Then he will not be able to say "donkey" when he sees a picture of a donkey, and the letter "d" will not help him either, because there are so many different words beginning with that letter. In fact, the "d" is more likely to activate "dog," since this word is more familiar, and the word "donkey" is not being activated by its picture because of the broken connection.

But here too the method of encouraging the patient to think of as many facts as he can about the animal in the picture can be useful, although for a different reason. In Carlton's case the animal-name network itself is intact. Thus, when he thinks of some facts about the donkey in the picture, his animal-fact network is activated, and this in turn activates his animal-name network. That is, thoughts about the animal in the picture eating hay and carrying loads may allow Carlton to think of the word "donkey" even though he cannot think of the word when he just looks at the picture. In this case the activation takes a detour around the damaged set of connections by using the intact connections going from the animal-picture network to the animal-fact network, and from there to the animal-name network.

These examples illustrate the way rehabilitation works after a stroke. In those cases where the rehabilitation is successful, its success is due to the strengthening of weak connections so as to form a detour around the damaged area. This also explains why so much practice is needed, and why the process is aptly described as "learning how to walk/talk/write all over again."

How is temporary memory different?

One mental function which has turned out to be practically impossible to recover after damage is the ability to store information in temporary memory. In the previous chapter we looked at various ways in which temporary memory differs from permanent memory. We saw that the networks which allow us to remember what we did five minutes ago and what we need to do five minutes from now are differently structured from the networks that let us name animals and think of many facts about them. Not enough is known yet about how the networks that serve temporary memory operate to be able to say why this function is so difficult to restore. But since the main mechanism for restoring function after damage in other areas of the brain seems to be the strengthening of connections that can constitute a detour, it seems plausible to suggest that there may be only one set of connections available for storing information in temporary memory, so that if these connections are damaged, no other connections can be recruited to detour around them.

Let us look at what actually happens in the case of damage to temporary memory. Let us say that Jane has had a stroke which also impaired her ability to move her left arm and leg, so that she is in the hospital for evaluation. She asks the doctor why she can't use her arm, and he explains the situation carefully. Jane seems to understand what the doctor is saying, so he goes on to evaluate the patient in the next bed. As he passes by Jane's bed again on his way out of the room five minutes later, she says, "Doctor, could you please tell me why I can't move my arm?" If the doctor is alert to the possibility of damage to temporary memory, he will realize that Jane may not have simply failed to understand his explanation; she may have totally forgotten that she just asked him this question and he just answered it. He will therefore order tests to examine the possibility that Jane's hippocampus, where her temporary memories should be stored, has been damaged.

What seems to have happened here is that the doctor's explanation was not encoded at all in Jane's brain. But let us take the story a bit farther. Jane's daughter, who is sitting with her in the hospital and has heard the doctor's explanation as well, tells it to her again and again, each time she asks. The next day, when the doctor comes back, he decides to test Jane's memory himself and asks her if she knows what is wrong with her arm. She answers that she had a stroke. He then asks her how she knows, but she cannot answer.

The reason that Jane can answer the first question but not the second is that her permanent memory has not been damaged. When her daughter answers her question about her disability again and again, the answer can be stored in whatever network is responsible for storing her knowledge about herself. This takes a long time, with many repetitions, because her temporary memory, which normally does the job of getting important facts into permanent memory networks, has been damaged. Eventually, however, as her daughter takes over the function of the temporary memory network by reminding her again and again of what has happened to her, this fact is stored in Jane's permanent memory. The memory about how she learned it, in contrast, is the sort of information that is generally stored only in temporary memory and not moved to permanent memory. Thus it does not find a place in permanent memory, and so is forgotten.

Trying to get information directly into the patient's permanent memory when her temporary memory is damaged can thus be seen as a sort of detour as well, since one form of memory is being used in place of another. The difference between this sort of detour and the one discussed earlier is that the first sort is a detour within the structures of the permanent memory networks, while the second involves the use of permanent memory in place of temporary memory, because of the impossibility of forming detours within temporary memory.

What sort of damage occurs in traffic accidents?

Traffic accidents typically cause a very different sort of brain damage than strokes. As we have seen, strokes tend to affect mainly the brain areas responsible for moving various parts of the body, speaking and understanding language, or some types of memory, as these are the areas most readily damaged by interference with their blood supply. Traffic accidents, in contrast, generally cause injury to the front of the head, as it impacts some hard, unyielding part of the vehicle. This results in damage to the frontal areas of the brain, the parts that are responsible for what are called "executive functions," such as setting long-term goals, seeking out problems that need to be solved, and planning how to solve them.

This sort of frontal brain damage can be hard to assess in the usual tests designed to measure intellectual performance in ordinary people, such as IQ tests. When patients with frontal damage are given these tests, they often do very well. This is because the areas of the brain that are involved in answering questions or solving problems once they have been presented are not damaged.

Let's say Zachary learned how to solve simple algebraic problems at the age of 13, and when he is 17 he is involved in a car accident that damages his frontal lobes. If the neuropsychologist testing Zachary's functioning after the accident gives him the sort of algebraic problem he knows how to solve, he may do very well, as the areas involved in solving such problems were not damaged in the accident.

Indeed, Zachary may have no trouble with any part of the IQ test, but if he is sent home he may simply stare at the wall or the television set all day without ever trying to do anything. This is because the part of his brain that has been damaged is the part that chooses what to do next and orders this choice to be carried out by the other parts of the brain, so he can't do anything on his own; he can only respond to the requests or orders of other people.

If Zachary's father, seeing this problem, then takes him to a rehabilitation center, they will have a hard job trying to help him. Such patients, in general, are much harder to rehabilitate than those who have lost the ability to walk or speak or write. This may be due to a problem similar to the one we discussed in the case of temporary memory. The executive functions may also be stored in only one set of connections, making it very difficult for the brain to set up a detour around them so as to regain its proper functioning.

Why should this be? Why should it be possible to relearn to use one's hands after a stroke, say, but not to plan one's future activities? Why should there be alternate, although little-used, pathways for writing, which can be strengthened by practice after the major pathway has been damaged, while there are no such alternate routes for planning one's activities?

It seems to me that the reason for this difference is the problem of interference. It doesn't hurt to have several different pathways for writing because they don't interfere with one another. I can print in upper and lower case letters, I can print all in capitals or all in lower case, I can write cursively in a very neat way, I can write very sloppily, I can write in large letters with chalk on a blackboard, I can write in tiny letters if all I have is a small piece of paper. At any one time I choose the way of writing appropriate to the current situation, and the fact that I have all these other alternatives does not interfere; I can just ignore them.

Thus if a stroke should damage the area which contains these writing pathways, there is a good chance that some of the pathways will be less damaged than others, since there are quite a few of them. Even if the main pathway is no longer usable, so that the patient cannot write at first, practice can strengthen one of the other, little-used pathways, to enable the recovering patient to learn how to write again.

But the situation is very different where planning one's activities is concerned. If there were alternate pathways for planning, I might plan two different acts for the same moment, and the result would be chaos. I'm not talking about the ordinary slipup where I make an appointment to go to the dentist on Tuesday at 10:00, and then, totally forgetting about it, make an appointment for the washing-machine repairman to come at the same time. This occurs because I wasn't thinking about the dentist appointment at the time when I made the appointment with the repairman. These are two activities which involve very different aspects of my life, and so I'm unlikely to be thinking about one when I'm thinking about the other. The two appointments are stored in two different parts of my temporary memory, each of which is connected with a different permanent memory network. Therefore there are no direct connections between them, and so outside help is necessary for co-ordinating them. Indeed, this is why appointment books (whether paper or electronic) were invented.

What I am trying to explain here is the need to avoid a situation which is hard even to imagine. Let's use a simple example. I am reaching for the coffee cup on the table, but the cup is behind my orange-juice glass, so I can't just pull the cup towards me unthinkingly, without any advance planning. Two possible ways for me to get the cup would be to reach around the glass and bring the cup to the front of the table in a circular motion, or to move the glass out of the way and then bring the cup directly towards myself. What is important here is not which choice I make, but that I should choose only one. If I had two parallel pathways for the process of choosing which one of these simple everyday acts to perform, then one pathway could be choosing one move while the other one was choosing the alternative move. Since these parallel pathways would be activated at the same time, they would send conflicting messages to my arm, and either I would end up doing nothing, or, more likely, I would knock over both the glass and the cup, so that the table, the floor and my clothes would be covered with orange juice and coffee.

The same problem would arise when I tried to write, if I had not only several writing styles to choose from but also more than one pathway for making the choice. It is certainly an advantage to be able to choose a small handwriting when I have nothing but a little piece of paper to write on, or a large handwriting to make the words on the blackboard visible to a large class. But if both alternatives were activated at the same time, I would be making the strokes for the large and the small letters in some random order, and the result would be totally illegible.

The pathways involved in choosing a career plan for the next stage of my life are undoubtedly much more complicated, but again, no matter how many alternatives I can *think* of, I must be able to *choose* one of them in the end. I cannot pursue a career as a doctor and as a lawyer at the same time; I have to choose and act upon one of these possible alternatives. This is probably the reason why there can be only one pathway for choosing which plan to carry out, and thus it can explain why there may be no way of compensating for it when it is damaged. There can be no detours here; there has to be one point where decisions are made, where action is initiated, and so if this pathway is damaged, nothing can replace it.

What happens in Alzheimer's disease?

The exact physiological processes that take place in Alzheimer's disease are still in dispute, but in general terms we can say that there are random disruptions of the connections in various networks in the brain. The tangles of neural axons and dendrites that are found upon post-mortem dissection of the brains of Alzheimer's patients are like tangles of wires in an electrical appliance. Just as an appliance cannot function properly if its wires are not laid out properly, so the brain cannot function if its neurons are not connected properly.

But this analogy is only partial. As we have seen throughout this book, the networks in the brain do not just send information from one place to another. All the information there is in the brain – not just knowledge about the world, but all our capacities to do the things that require a brain, from walking upright to going shopping to cooking meals to discussing philosophy over dinner – is located within the networks themselves. An accumulation of random disruptions of the connections in any network will lead to increasingly erratic behavior of various sorts, ending in the inability to do anything at all.

The tangles in the networks that constitute Alzheimer's are quite different from the ordinary death of neurons. We have seen that we are constantly losing neurons, but that the loss of a few of them does not substantially change the functioning of any given network. In normal aging many neurons die, and this may be one reason why it becomes harder to learn new things as we get older. But as long as the remaining neurons retain their proper connections, we can still do all the things we could do before, even if it may take a little longer.

The way Alzheimer's differs from normal aging is that the neurons don't just die and disappear – instead, the axons and dendrites that connect them become tangled up. In normal aging, some of the connections that have activated the pattern yielding the name "Judith" when I see my friend may be lost, and so it may take a little longer than usual for me to recall her name. The remaining connections will still be intact, however, so I will not mistake her for someone else. But if John, say, has Alzheimer's, then the axon branches coming from the human-face network that are supposed to activate some dendrites in the human-name network to allow John to say "Jerry" when he sees his friend may become tangled up with other axon branches in the face network. This may lead to a situation in which several different patterns in the name network are slightly activated at the same time, with no one of them being activated sufficiently to allow the neurons to form one stable pattern for one particular name. John will thus be unable to recall Jerry's name.

As the tangling of the dendrites and the axon branches spreads to other areas of John's brain, he may no longer recognize Jerry as his friend at all. Later, even though he may still be able to talk, he may not recognize his own children when they come to visit him. They may seem familiar, but he will not remember how they are related to him. Later still he will not be able to perform even the simplest functions.

The loss of function in patients with Alzheimer's does not always occur in the same order, as the random tangles may begin in different areas of the brain and spread in different ways, but in the end the entire brain is affected.

Can anything be done about Alzheimer's disease?

The questions of interest to most of us are whether Alzheimer's can be prevented, whether it can be arrested once it starts, and whether the damage it has caused can be reversed. Prevention is theoretically possible, once we understand how the disease comes about, and this understanding may also make it possible to stop the disease at any point. But reversing the damage once it has occurred is not even theoretically possible. There is a great deal of discussion at present of cures for other diseases of or injuries to the nervous system, such as Parkinson's disease or spinal cord injuries, but it seems to me that none of these cures would be able to reverse the damage in Alzheimer's once it has occurred.

Until very recently the accepted wisdom in biology was that nerves in the central nervous system, which includes the brain and the spinal cord, can never regenerate. That is, once they are damaged, the damage is permanent – the damaged cells cannot repair themselves, and the undamaged cells cannot divide to provide new cells.

Our skin, for example, repairs itself after injury in that the undamaged cells divide rapidly to provide new healthy cells that take the place of the damaged ones. In cases where so much skin is damaged that this process cannot provide enough new cells, skin can be taken from another part of the body to cover the injured area, and the cells in this graft will divide in their new home and connect up with the ones already there. This is a relatively easy process because skin cells are connected to their neighbors very simply – all they have to do is touch each other snugly, as their main function is to protect the body parts underneath them.

Nerve cells, in contrast, generally stop dividing around the time of birth. Learning, as we have seen, takes place through the selective strengthening of connections between neurons, and does not require the growth of new cells. And until very recently it was believed that the neurons cannot be stimulated to divide again, even in the case of injury.

In the past few years, however, there have been hints that a way can be found to make the neurons in the spinal cord start dividing again after an injury. This technique is still in the laboratory stage, and has not yet been refined enough to start using it in injured people, but let us assume that it will someday be used in patients. Such an advance might be possible because the nerve cells in the spinal cord are connected to one another in fairly fixed ways. These neurons are merely messengers that allow the brain to tell the muscles what to do. They normally form connections with one another before birth, and these connections are not changed by the learning process. When we learn to walk, say, changes occur in neural connections in the motor networks of the brain and in the cerebellum, the area at the base of the brain responsible for fine-tuning our movements. The connections between the neurons in the spinal cord, however, remain fixed. Thus if there is some way to get these neurons to regenerate after they have been damaged, it is plausible that they might be able to reconnect to one another in the way they did when they were first developing, and this may be enough to allow the patient to walk once again.

The situation in the brain, though, as we have seen throughout the book, is far more complex. The connections between the neurons are changed by everything we learn during our lives, and these changed connections actually *constitute* our knowledge. Thus if they become tangled up, there is no way to untangle them. Even if the remaining healthy neurons could be made to divide by some spectacular medical breakthrough, the odds would be astronomical against any new connections they made being the right ones. Since these connections were all established by a slow learning process in the first place, any reconnection would require a similar slow learning process.

Indeed, the evidence of occasional recovery from serious brain damage offers strong support for this claim. Some people who have been in a coma for a long time, generally after suffering brain damage due to the brain being starved of oxygen for too many minutes, do eventually regain near-normal functioning. But this requires a lengthy learning period, in which the patient must learn to walk, talk, feed and dress himself, all over again. In such cases many of the connections have apparently remained intact while others were damaged, so that the patient's expert knowledge in a particular field may be preserved and become available once again when the patient becomes able to access it.

Thus if some way were found of making the neurons in the brain divide, the most that could be hoped for in the case of Alzheimer's patients would be that they could start the process of learning all over again. But since in the case of Alzheimer's, unlike the case of oxygen starvation, the patient's knowledge about the world is often one of the first casualties of the disease, it would probably take as many years to relearn all this knowledge as it took to learn it in the first place. And since even the brains of normally aging individuals have trouble making the changes that are required for learning a great deal of new material, the prospect of restoring Alzheimer's patients to their former situation seems extremely remote.

Our only hope, then, is prevention, which remains possible even if a cure may not be. But this is not a forlorn hope. Mankind has practically eradicated many incurable viral diseases, such as smallpox, through prevention, so we may hope that a preventive measure will be found in the case of Alzheimer's as well.

11

Practical implications

Is there any way that all this new knowledge can help us in our daily life? In addition, is it possible that it might be of some use to psychologists in helping their clients?

Since we have spent most of the book starting with questions and looking for answers, let us try it the other way around now: Let us start with the answers and see what questions they can answer, as on the old television quiz show *Jeopardy*, which has recently been revived. There is a good reason for this, aside from the obvious one that I can avoid those questions I have no answers for. The deeper reason is that there are some questions that most of us don't think of asking at all, and starting with the answers may lead to some of these less obvious questions.

How to study more efficiently

The first answer is that the knowledge structures in our inner networks change very gradually when we provide them with new information. One obvious question to which this may provide an answer is "What is the best way to study new material?" What we have learned in this book is that our mental networks cannot handle large amounts of new information at once because they can make only small changes in the way our knowledge is organized. Thus the best way to learn new information is to study small amounts at a time and keep trying to think of ways to integrate it with what we already know.

Of course, students have always been told to spread out their studying throughout the year and avoid cramming for tests, yet they continue to do most of their studying the day before the exam. The distinction between the two different types of memory described in the last chapter offers an

"Who was the first president, Jack?"

explanation for why cramming seems to work, thus reinforcing students' tendency to use this inefficient means of study. Cramming saturates temporary memory with a great deal of information, which can be remembered for a day or two just the way we can remember many of the details of what we did today for another day or two. But just as these details are soon lost because they have not been stored in permanent memory, so the information we stored temporarily while cramming is soon gone, and when we get to Advanced Physics we find it hard to remember the material from Physics 101.

One way of studying that I find extremely useful is to read a little of the material for one class, then take a little break, thinking about this material and how it relates to what I already know, then read a small amount of material on a different subject, then think about that for a bit while doing some boring chores, then go back to the first subject, and so on. This method is much more efficient than it seems because the different subjects are stored in different networks, so while the first one is adjusting to the new information it has just received and would thus have difficulty processing additional quantities of information, the other network has not had to change its structure for a while and so should have less difficulty absorbing new knowledge.

Another study tip that takes advantage of the fact that our networks change their structure only gradually may be called the "overnight break." I have found that when I have to read a great deal of difficult new material in a short time, say in the week between one class and the next, it often happens that I simply do not understand what I am reading, even if I take short breaks in between. Eventually I become so frustrated that I just stop, resolving to try again the next day. And lo and behold, when I read the same material over again the following day it is much clearer – not necessarily easy, but at least comprehensible.

The reason for this lies again in the interplay between temporary and permanent memory. The material I read the first day could only be absorbed into my temporary memory, as it required large changes in the connections between the neurons, which the permanent networks cannot handle. But the material could not be *understood* in the temporary networks, because these networks do not contain the permanent structures within which the information could be integrated. These structures exist only in the permanent networks, which need to change gradually in order to integrate the new information. During the sleep period between one study session and the next, the information stored without comprehension in the temporary network gradually changes the structure of the appropriate permanent network so that a basis for comprehension is established, and when I reread the material the next day I can fit it into my permanent knowledge structures, which is the definition of understanding.

How to teach more efficiently

Obviously, one way to teach more efficiently is to encourage students to use the study methods outlined above. But another answer that emerges from the discussions in this book is that it is difficult to learn anything completely new, because our minds rarely form new networks from scratch. What they generally do is change old ones, sometimes developing them in such new ways that in the end a new network exists. But each new network is based initially on some old one, even if it may eventually become independent of the previous one.

This knowledge can help us answer the important question, "How can we teach our children and our students in such a way that what we are teaching them becomes a part of them, not something imposed from the outside?" Or, to use a popular psychological term, "How can we get others to internalize what we teach them?"

What we must keep in mind are the implications of the fact that no one can learn anything completely new. This means that anything we teach must link up with something that the learner already knows, elaborating it or changing it in some way. Whether we are parents of a young child or teachers of older children or adults, we must try to figure out what they already know and add the new knowledge to the old.

How does this work? With little children, most parents are pretty well aware of how much the child knows. If we want to teach two-year-old Emily not to hit her friends, we can say, "You know how much it hurts you when someone hits you. Well, it hurts Pam the same way when you hit her. That's why she's crying, just like you when you get hurt. So don't hit her." The network that stores Emily's awareness of her own feelings can then change slightly to include the awareness that other people have similar feelings.

Just saying "Don't ever hit anyone" isn't very helpful because there isn't any old knowledge for it to be attached to. All it can be connected with is the presence of the parent who is saying it, which may explain why children who are ordered to do things without explanation are likely to obey only when the parent or teacher is around. The rule remains outside the child's own thinking, connected only with what seems like the parent's arbitrary command.

In contrast, the child who is given an explanation that ties in with what she already knows can internalize the command. She can learn to sense her friend's pain in a way that is connected with her awareness of her own feelings of pain. She can therefore begin to feel that she herself does not want to hurt her friend, not merely that some authority has made a rule to this effect. Such learning is deep and long-lasting because it is anchored in already-present knowledge.

This is, of course, a very simple example, one that many parents use naturally. How can this technique be applied to teaching really difficult concepts, such as atomic physics?

There are two important ideas that can be helpful here. One is to continue the use of analogies. Just as we help children understand how others feel by showing how the feelings of others are similar to their own, we help students of all ages understand the structure of the atom by drawing an analogy with the structure of the solar system. This anchors the new information about the atom in the old information about the solar system, setting up similar patterns of activity associated with the new concepts. Once the structure is in place and the new names are learned, we can then proceed to the next step and point out the differences between atoms and solar systems. This may eventually result in two separate information structures, one for the solar system and one for the atom, but it would have been much harder to develop the new one without using the old one.

To be sure, there are problems with this technique. Since we are starting with an analogy, there is always the risk that the students will carry over some inappropriate aspects of the solar system to the atom. This risk, however, is less than that of the lack of understanding if the structure of the atom is presented without any analogy to more familiar structures. It is better to use the analogy and then refine the differences later than to try to avoid analogy altogether.

There is also another technique that can be used with older children and adults, especially if you are a teacher faced with a classroom full of students whom you do not know well. This technique is meant to deal with the problem that you cannot connect the new information you are trying to teach with the information the students already possess if you have no way of knowing how much they actually know.

The idea is to elicit feedback from the students that will reveal what they know about the topic and what they do not know. This is not meant to be an invitation for a free discussion in which students express their opinions on topics about which they know very little. The idea is to begin by presenting some interesting information about the topic that you want the students to learn. The students should then be encouraged to ask questions about it, to tell the class what it reminds them of, to present any information about it that they already have.

You, as the teacher, can then use this opportunity to connect the new material on this topic with the old information. The students' questions and comments will let you know how new the topic is for them. If they offer comments which show that they have already learned a great deal about the topic, you can show them how the new information is an extension of what they know. If, in contrast, they ask questions which show that the topic is totally new to them, you can backtrack and explain the new information more simply, until you find some way of connecting it with something they do know about. In this way the students can form long-lasting connections between the old and the new knowledge.

How to overcome unhealthy patterns

Another important answer we have come across is that we have "attractors" in our mental networks. That is, there are some patterns of activity that have become so greatly strengthened by repeated use that any new incoming information that sufficiently resembles our old knowledge is treated as if it were exactly the same as what we already know. We used this fact (in Chapter 6) to answer the question, "Why do we treat members of not-very-familiar groups as if they were all the same?" But it can also provide an answer for the question, "Why do we react to events of a certain type as if they were the same as previously occurring events of that type?"

This last question is one that is of great interest to therapists. People in therapy often complain that the same bad thing keeps happening to them over and over – for example, they are constantly being fired from each new job they manage to find. The therapist may suggest that the client sees each boss as if he or she were the same as one of the client's parents. Let's say Joe is the client and his boss is Wanda. Even if Wanda does not resemble Joe's mother in any way, the mere fact that she is in a position of authority over him makes the situation similar to ones he was often faced with when his mother had such authority. Joe therefore reacts to Wanda in the same rebellious ways he used towards his mother, and she counter-reacts by firing him.

Can connectionist theory provide us with a way for therapists to help clients with such a problem? Let us recall how we can overcome the pull of an attractor – say, how we teach a child that not all four-legged animals are horses. One way we can do this is to show the child pictures of the different animals and point out the differences between them while giving each one a different label. For example, we say, "No, that's not a horse, it's a cow. See, it doesn't have a mane, it has horns, its shape is different." This enables the child's animal network to form a different pattern of activity connected with the word "cow" in the animal-name network that is not automatically pulled into the pattern connected with "horse." It is repeated emphasis on the differences rather than the similarities that makes it possible to form new patterns of activity in the networks, thus allowing the global concept to be differentiated into more specific concepts.

The same process can be applied in the case of Joe. What Bill, his therapist, tries to teach him is that each specific job and each specific boss is different. Bill asks Joe to report how he reacted to a particular act of Wanda's, and then tries to get him to see the differences between the situation on the job and earlier situations at home with his mother. Seeing the differences enables Joe to form new patterns of activity for reacting to his boss's demands and keep these patterns separate from his old patterns of reacting to his mother's demands.

Another way of getting out of "attractor" patterns is by a radical shift in perspective. One perspective shift often suggested by family therapists is role reversal. This means that two people involved in a conflict, say mother and daughter, each plays the role of the other, with the mother presenting arguments supporting the daughter's desire for independence and the daughter suggesting reasons why she should be the kind of person her mother wants her to be. This technique often works unbelievably well, resolving conflicts that have lasted for months or even years. Can connectionist theory explain this success?

I have not found any explanations for the effectiveness of perspective shift among academic connectionists, but an idea which pervades the writings of Edward deBono seems to me to provide such an explanation. In Chapter 5 I mentioned deBono as one of the precursors of connectionist theory, sadly ignored by the establishment. DeBono emphasizes in many of his writings that when there are attractor patterns in the mind, our thoughts keep going around in their well-established circuits, preventing us from thinking of any new idea that might provide a way out of some persistent dilemma or conflict. DeBono offers his solutions to this problem mainly for business conflicts, but it seems to me that these solutions should be applicable to therapy settings as well.

What deBono suggests is that any familiar idea will be immediately "captured" by some attractor pattern, so the only way to get out of this rut is by deliberately introducing an unfamiliar idea, one that does not fit in with the notions already possessed by the person. This idea, being unfamiliar, does not resemble any of the attractor patterns, and so will not be captured by any of them. Instead, it will form a new pattern of activity, as described in Chapter 6, with the new pattern on the same network as the old ones, since it involves the same basic topic. The new pattern, being surprising and therefore attention-grabbing, will be able to influence and change the old ones instead of being changed by them.

DeBono offers examples of outrageous assertions to be used for this purpose, such as "Let's say cars would have square wheels," to be used to help develop new ideas for the design of new cars. But it seems to me that taking the perspective of the other person is unfamiliar enough to people caught in a conflict for it to work similarly. Thinking about her teenage daughter's need for independence, instead of her own wish to mold her daughter into the kind of person she thinks the girl should be, is a good way to get the mother's thoughts out of the well-worn groove of how to influence her daughter. Instead, she could try to think of constructive ways of helping her daughter achieve the independence she wants without totally abandoning the family's values.

Conclusion

These are only a sample of the many possibilities for understanding and possibly changing our ways of thinking about important issues by using the knowledge about our mind provided by connectionist theory. You too can use the technique of formulating a question for every answer instead of merely trying to think of answers for questions. Actually, this answer-questioning technique is yet another example of using deBono's method of challenging the familiar by the use of the unfamiliar.

Using the unfamiliar does not always lead to abandoning the familiar. The act of questioning our familiar answers may lead to a deepened appreciation of the old answers as often as it leads us to formulate new ones. But either way, we will no longer be following along an old path merely out of habit, merely because we have never taken the trouble to examine it. Understanding how our mind works provides us with the opportunity to do what we do deliberately, out of choice. Thus seeing how the mind's functioning is grounded in the brain's activity does not make us more mechanical, more computer-like; on the contrary, this self-reflection frees us to be more human.

12

Criticism of connectionist theory

Until now I've been presenting the connectionist viewpoint of the way our minds work as if it were generally accepted. This is because, as I noted at the outset, it would have been far too annoying to repeat "according to the connectionists" in every paragraph. But actually this approach has come in for its share of criticism. Well, then, what sorts of criticisms are these? And how do connectionists respond to them?

The representationalists' criticism

Probably the most important criticism comes from a group of philosophers of science who espouse a theory known as "representationalism." This theory is one of the "classical" theories I mentioned in the Introduction, and two of its most prominent advocates are Jerry Fodor and Zenon Pylyshyn. As an alternative theory of mental processes, it too is complicated and cannot be explained in a few sentences any more than connectionism can. What I will try to do, then, is to point out some of the ways in which representationalism differs from connectionism, so that you can get some idea of what is at stake in the debate between the two camps.

One of the important differences between the two theories is the one that gives them their names. Representationalists claim that every object we see, every sound we hear, every word we learn, is stored in our mind as a separate, point-like representation of that thing. Just as the word *cat* represents cats, they say, there is a concept CAT inside our mind that represents, somehow, the essence of "cathood." In the connectionist view, we recall, our notion of a cat is distributed over several networks, and is related to other associated notions such as "dog" or "fur" by sharing part of the same networks or by activating particular patterns on some other networks. The representationalist view, in contrast, claims that our concept CAT is a particular point which is related to other concepts through logical, structural relations. For example, CAT is related to DOG through the facts that "A cat is a mammal" and "A dog is a mammal."

This should sound familiar by now. It should, indeed, call to mind the description of semantic networks in Chapter 4. And for good reason, since semantic network theory is a typical representationalist theory. It is not the only one, since some representationalists believe, for example, that there are lists of concepts stored in our head somehow. All of them, however, insist that these concepts are stored as simple units that are then combined to form propositions, which are the ideas underlying sentences.

Moreover, Fodor and Pylyshyn claim that these combinations are systematic, that if we have learned, for example, that "The boy loves the girl" is an acceptable sentence in English, then we also know that "The girl loves the boy" must be acceptable, without having to learn that fact separately. They are not saying, of course, that if one of these sentences is true then the other one also has to be true; most of us know only too well how often one of them can be true while the other one is false. What they are saying is simply that if you can understand and use one of these sentences, you can also understand and use the other, without having to learn it separately.

The reason they give for this is that the sentences are built up out of their individual parts, so these parts can be combined in various ways. In their view, parts of the same kind, such as "the boy" and "the girl," are therefore interchangeable in their ability to form possible sentences of English. Thus, they claim, any cognitive system must be able to interchange such elements, and a representationalist system can do this while a connectionist one cannot.

Answering the representationalists' criticism

There are two different ways to respond to Fodor and Pylyshyn's criticism. One is to point out that there is a great deal of evidence that the human cognitive system does not behave the way they describe it. The other is to show that, in those cases where their description of the way we think does seem to be fairly accurate, connectionist systems can be developed to perform these tasks as well.

An example of the first type is Fodor and Pylyshyn's insistence that parts of sentences must be interchangable. Here we face the question of what is meant by "the same kind of part." Now only nouns or noun phrases can be put in the positions of A and B in the sentence "A loves B." But can all nouns be put into these positions? If I can understand "The boy loves the girl," does this mean that I can understand "The house loves jumping"? Yet "jumping" can be used as a noun in certain sentences quite interchangeably with ordinary nouns like "bicycle." Children who can understand "No bicycles in the house" can also understand "No jumping in the house." Yet they also know that "The boy rides the bicycle" is an acceptable sentence, but "The boy rides the jumping" is not, and neither is "The bicycle rides the boy." But "The bicycle rides the boy" is the reverse of "The boy rides the bicycle" in exactly the same way as "The girl loves the boy" is the reverse of "The boy loves the girl." Some people eventually come to accept that "The bicycle rides the boy" is a grammatically well-formed sentence even though it makes no sense, but it is certainly possible to be considered a speaker of English even if one doesn't accept this.

Thus Fodor and Pylyshyn are wrong in their claim that parts of sentences are interchangeable in some simple way. On the contrary, the way we learn which words can be substituted for other words in particular sentences is very complex. But since some types of words can certainly be substituted for others to make acceptable sentences, and this knowledge is basic to human language, connectionist theories do have to be able to model this ability.

This brings us to the second type of response. In addition to insisting that human thought is not as systematic as the representationalists claim, connectionists can respond with attempts to model those aspects of human thinking that *are* systematic. Although connectionist theorists do not yet have a full explanation of how these complex cognitive tasks are performed, at least a start has been made. In Chapter 7 we saw how relational networks can store sentences such as "Dinah is the daughter of Leah." These sentences are connected to the words they contain – for example, the sentence just mentioned is connected to the word "daughter" in the family-relation network. Since "daughter" and "mother" have a great many activation units in common in this network, they are very closely associated. Thus we also accept "Leah is the daughter of Dinah" as a proper sentence of English, even though we know that both of these sentences cannot be true of the same two women.

The mutual connections between sentence networks and word networks can also begin to provide an explanation for the fact that we do not conclude that "The boy rides the jumping" is an acceptable sentence even though "No jumping in the house" is perfectly acceptable. "Jumping" and "bicycle" are not found in the same word networks. "Jumping" is in a network of action verbs, together with "running" and "walking" and "climbing," while "bicycle" is in a network of artifacts, together with "trains" and "airplanes." Thus "jumping" and "bicycle" have no activation units in common, and very few connections between them. No one would ever think of substituting one for the other on general principles. The only reason we accept the sentence "No jumping in the house" is that we have heard either this sentence or sentences with other action verbs such as "No running in the house." We could not simply predict that it was acceptable on the basis of knowing that "No bicycles in the house" is acceptable.

Thus, even though connectionists cannot yet completely explain how we know which words fit into which sentences, neither can their opponents. Moreover, the representationalists don't even seem to realize that there is a problem here that needs explaining, so they certainly can't deal with it. Thus this particular critique of connectionism does not seem to be justified.

The claim that connectionism is unscientific

Although Fodor and Pylyshyn believe that connectionist theory is wrong, they take it seriously and present reasoned arguments against it. In the previous section I summed up one of their key arguments and explained why I think it is not a valid criticism. But this is a fair argument on both sides, with mutual respect. Quite a different sort of argument is presented by A.K. Dewdney. Instead of taking connectionism seriously, he tries to dismiss it as what he calls "bad science." If his "arguments" had been presented by a person lacking academic credentials, I myself would not take them seriously enough to try to rebut them. But Dewdney is a professor of mathematics and former writer of the column on mathematical recreations for *Scientific American*, and it is quite easy to be taken in by his ridicule. Moreover, some of the arguments he presents have been put forward by other critics of connectionism. It may therefore be useful to go through them one by one.

Dewdney's arguments against connectionism are presented in a book called *Yes, We Have No Neutrons: An eye-opening tour through the twists and turns of bad science.* In the introduction to this book he sets forth his definition of good science and bad science, and he explicitly claims, "Every piece of bad science in this book may be traced back to a particular error of method." The rest of the book contains examples of what he considers to be bad science. In most of these cases Dewdney applies his guidelines and specifies the particular errors perpetrated by the researchers. In the case of connectionism, however, he abandons his own guidelines and puts together a variety of different types of criticism that may at most show that certain applications of connectionism are misguided and that certain claims of connectionists may be wrong.

To be wrong, however, is not to be unscientific. It seems to me that Dewdney has some sort of unreasoned prejudice against connectionists which leads him to confuse a struggling new science, which is often confronted by setbacks and has to rethink its hypotheses, with bad science. Using Dewdney's own guidelines for good and bad science, I will try to show why I believe he is wrong in the case of connectionism even though he is apparently right in most of the other cases he describes in his book.

First I will present Dewdney's guidelines. For each one I will show (a) how it is followed by "good" scientists; and (b) how it is flouted by the researchers in Dewdney's examples of "bad" science; then (c) I will examine the procedures of connectionists to see whether they follow the guideline or not. Finally, I will go through Dewdney's specific criticisms of connectionism, one by one, and show why they are either wrong or irrelevant.

Dewdney's guidelines for good science

(1) Formulate an interesting question that involves a general law.

(a) Newton, for example, asked what mathematical law might govern the process of bodies falling towards the earth, and whether the same law could predict the motions of the planets around the sun. This is certainly an interesting question, and in attempting to show that the same law applies on earth and in the "heavens" Newton was striving at a generalization unusual before his time.

(b) When Binet developed the first IQ tests, his only question was whether certain children who seemed unintelligent might actually be able to benefit from schooling. This is not a scientific question but a practical one. Binet himself never claimed it was a scientific issue, but later researchers did, even though they never actually formulated any scientific question. Measuring children's IQ may be helpful for deciding how to teach them, just as measuring their foot size is helpful for determining what shoes they should wear, but neither process is a part of science.

(c) Connectionists have asked many interesting scientific questions. For example, are concepts represented in our brains individually in separate units or in an overlapping fashion, over all the units of a network? This and other questions asked by connectionists are meant to be as general as possible, applying to all terrestrial creatures with brains large enough to have such a complex structure.

(2) Formulate a hypothesis that can be tested to see if it is true or false.

(a) Mendel formulated the hypothesis that there are dominant and recessive genes. If the gene for long-stemmed pea plants (L) is dominant and the gene for short-stemmed ones is recessive (S), and every plant has two such genes, then two pea plants that are long-stemmed in appearance but actually have one of each gene (LS) will have offspring three-quarters of which are long-stemmed (LL, LS and SL) and one-quarter short-stemmed (SS). If the actual ratio in many such experiments had been very different from 3:1, Mendel would have been the first to agree that his hypothesis was false.

(b) Freud formulated the hypothesis that unconscious thoughts, beliefs and desires motivate much of our behavior, but he did not design any tests to find out if this hypothesis is true or false. On the contrary, he built into the method of psychoanalysis a way of preventing anyone from ever showing that his hypothesis is false. He claimed that we all have resistances to seeing the workings of our unconscious, so whenever we bring any evidence that would seem to contradict the existence of some hypothesized unconscious force, this is merely the product of our unconscious resistance to the truth. Since Freud did not provide any criteria for what would count as falsifying his hypothesis, it is not scientific.

(c) Connectionists have always tried to formulate their hypotheses in ways that can be tested. For example, Rumelhart and McClelland formulated the hypothesis that our perception of letters depends not only on the lines and curves making up the letters but also on the words in which the letters are embedded (as described in Chapter 8). If it had turned out that their computer model of the way the perception of words affects the perception of the letters in the words did not match the results of experiments showing that human beings perceive letters better within words than in isolation, then Rumelhart and McClelland would have abandoned their hypothesis and tried to formulate a new one.

(3) Design an experiment to test the hypothesis, perform the experiment and record the results.

(a) Galileo designed the experiment of rolling objects down an inclined plane to test his hypothesis that objects of different weights fall at the same speed. He performed the experiment and carefully recorded the times it took for each of the objects to go from the top of the inclined plane to the bottom.

(b) Freud never designed any empirical studies to test his hypothesis that traumas cause neurotic symptoms. Obviously, it would have been unethical to perform experiments in which he inflicted traumas on children to see if they developed neurotic symptoms as adults. He could, however, have designed an observational study to see if people with neurotic symptoms treated by his method were more likely to get better than people with similar symptoms treated by other methods used at that time, or people who received no treatment at all, but he did not do so.

(c) Connectionists set up computer models to see if their hypotheses about mental processes are supported by the results. Rumelhart and McClelland designed a computer model of how letters are perceived within words. They then compared the pattern of results produced by the computer model with the pattern of results produced by human subjects faced with the same task. This is an empirical test of the correctness of a theory similar to those used by scientists in other fields.

(4) Work in teams and share one's results with the scientific community so that errors will be discovered quickly and fruitful results will accumulate into a body of knowledge.

(a) Watson and Crick published their results about the helical structure of DNA, and these were soon examined by many other molecular biologists. Different teams set about determining the structure of the genes in different organisms, correcting each other's errors and accumulating a large amount of knowledge about the structure of many different genes.

(b) Fleischmann and Pons were originally reluctant to publish any details about their "cold fusion" experiments because they were afraid of losing the advantage of being the only ones to know how to do it. When they did publish some details and others tried to repeat the experiments, they did not get the same results. The whole enterprise bogged down in trying over and over to repeat the original results and did not lead to any accumulation of new knowledge.

(c) Connectionists prefer to work in teams so as to formulate and test models of all sorts of different mental functions. Rumelhart and McClelland formed the "PDP Research Group" to investigate various aspects of "parallel distributed processing," the type of connectionism described in this book. Different members of the group are constantly scrutinizing each other's work, partly to avoid errors, but mainly to be able to build on each other's results, to accumulate a body of knowledge that no one individual or pair of individuals could achieve on their own.

Dewdney's guidelines for bad science

So far we have seen how connectionism conforms to four guidelines for good science set forth by Dewdney. But Dewdney also describes two pitfalls of bad science. Could his aversion to connectionism be based on its succumbing to one or both of these pitfalls? Let us see.

The first pitfall Dewdney mentions is that bad science occurs when researchers believe they have made such a crucial, earth-shaking discovery that they must proclaim it to the popular media before going through the process of publishing their results in scientific journals. In such journals research is reviewed by one or more respected scientists in the field, precisely to make sure that the guidelines for good science have been followed, and that the results are important enough to be worth publishing.

Clearly connectionists have not succumbed to this pitfall. None of their individual results has been exciting enough to warrant a press conference; it takes long, careful study to appreciate how important these ideas are. Connectionist research is published in journals of the same type and quality as those that publish the results of other research in cognitive psychology.

The second pitfall is the refusal to admit that one might have made a mistake, that all of one's results are based on some crucial error. This pitfall is the flip side of guideline no. 4 above. It happens only when researchers work alone, without input from their peers. Since, as I have shown, connectionists carefully follow guideline no. 4, they do not have to worry about this pitfall. If they make mistakes, there are plenty of people around to tell them to try something else.

Dewdney's specific criticisms of connectionism

The difficulty in responding to Dewdney's criticisms of connectionism is not in finding answers to them but in finding the criticisms themselves. Since he couches them in a tone of selective ridicule, where, for example, connectionists are described as "starry-eyed" while their critics are described as "painstaking," the first and most difficult task is to separate out the real criticisms from the ridiculing adjectives. Once this is done, it is not too difficult to answer them.

(1) The perceptron

The first criticism is of a device called a "perceptron." This was a device, invented by Frank Rosenblatt in 1962, which consisted of a single layer of neuron-like units, similar to the ones described in this book, that was capable of performing simple logical tasks and even learning to do new tasks. But in 1968 Marvin Minsky and Seymour Papert, two advocates of artificial intelligence, showed that there are some logical tasks that perceptrons are incapable of ever learning.

This criticism held back the development of connectionism for a few years, but eventually it was discovered that if several layers of neuron-like units are used, forming networks similar to those discussed throughout this book, the limitations of single-layer devices could be overcome.

Thus to criticize neural networks because they are made up of units each of which is incapable separately of performing logical tasks is like saying that, since you can't get very far on a unicycle, a bicycle can't take you very far either.

(2) Applications of neural networks

Alongside the connectionists, who are trying to create a model of the functioning of the human mind based on the proven scientific fact that the human brain consists of networks of neurons, there are computer scientists who are trying to create artificial neural networks to solve problems on computers. Dewdney claims that these artificial networks have not been able to solve all the problems that their proponents have designed them for.

Here it is totally irrelevant for our purposes whether Dewdney's criticism is correct or not. Even if Dewdney is right and artificial neural networks will never be able to solve all the problems that human beings can solve, this does not prove that the real neural networks inside human heads do not solve problems. They obviously do so, since human beings have nothing but the neural networks in their brains to solve problems with. The question of whether artificial neural networks can solve problems is similar to the question of how long people can be kept alive on an artificial heart or kidney built to resemble the human heart or kidney. This question is irrelevant to the question of whether the theory about the functioning of the human heart and kidney that was used as the basis for building the artificial ones is a correct theory of the human organs themselves. In the same way, the question of whether artificial neural networks have yet been invented that can solve some particular problems is irrelevant to the question of whether neural network models are the best way of accounting for the human ability to solve these problems.

(3) The hill-climbing problem

Here we have an interesting, relevant problem. One of the implications of the way neural networks are constructed is that they can be changed only in very small steps. In Chapter 6 we saw that learning takes place when very small changes are made in the connections between the various neurons in a network. Once the changes have produced a pattern that is sufficient to solve some particular problem, the network sees no need for any more changes, and the problem-solving process stops. The particular solution that has been arrived at becomes an attractor, and whenever a similar problem is posed, the network will automatically produce the same solution it gave for the previous problem.

Unfortunately, the solution to the old problem is not always the best solution to the new one, even if the problems seem similar at first glance. Once a particular solution has become an attractor, however, it will impose itself on similar problems, even if there might be a better way of solving these problems, just as a child's animal-name network may lead her to say "dog" when she sees a cat for the first time.

This problem is called the hill-climbing problem because it resembles the problem of trying to climb a mountain in a fog. Since you cannot see the top of the mountain, all you can do is follow the rule to keep moving to higher ground. The problem is that once you have reached a local peak, every path away from it will be downward, and so you will stop climbing, even though the mountaintop may actually be very far away.

For computer scientists who are trying to find the optimal solution to a problem, this is indeed a snag. The question for researchers who want to know how the human mind functions is, however, quite different. We are interested in knowing whether this is actually what the mind does, and the answer is that this is exactly how it operates. Both everyday observation and careful laboratory experiments show clearly that once people have found a good-enough procedure for solving a problem, they generally continue to use this procedure for similar problems and do not try to search for a better way of solving them.

Thus this particular "failing" of neural network models, like many others, is actually another piece of evidence that these models capture important aspects of how the human mind functions. What is a failing when you are trying to build a perfect mechanism turns out to be an advantage when you are trying to understand how an imperfect mechanism works.

(4) Do the neuron-like units of network models resemble real neurons?

Another criticism that has been leveled against network models is that the units in these models are not very similar to real neurons. I'm not sure why this is considered a criticism. At the beginning of this book I explained that connectionists trying to base their models on brain function are not committed to a particular neural unit in the brain as the analogue of the neural unit in the model. They simply chose the neuron as the basic unit to model because it is the most plausible candidate on the basis of our current knowledge. If it later turns out that smaller parts of the neuron are actually the basic units, or if, on the contrary, we find that assemblies of several neurons are the basic units, the units in the model can be seen as analogous to them. Thus the lack of precise similarity between the biological neuron and the unit in the models is actually an advantage rather than a disadvantage.

(5) The past-tense model

One criticism that is very important to deal with is the criticism of Rumelhart and McClelland's model of how children learn the past tense of English verbs (one of the models described in Chapter 8). This model has been faulted for failing to conform to the way human children actually perform in a number of respects. For example, when the computer program was confronted with several irregular verbs it had not encountered before, it produced the correct past tense for only four out of eleven verbs.

Perhaps the most important criticism of the model is that of Steven

Pinker and Alan Prince, who are followers of the representationalist theory described at the beginning of this chapter. Pinker and Prince believe that children learn the past tense by figuring out the rules on which the past tenses of most verbs are based and memorizing the exceptions separately. They list a number of failings of the Rumelhart and McClelland model, claiming, for example, that it cannot learn certain rules that are prevalent in human languages, yet is capable of learning some rules that are not found in any human languages. They use these failings as evidence that the rule-following theory is the correct one and that connectionist models of learning without following rules do not fit the facts about how children actually learn.

A carefully thought-out response to this criticism appears in a paper by the connectionist Mark Seidenberg. Seidenberg claims that the essence of the debate between representationalists and connectionists on the issue of language learning is captured by one particular controversy between them: Representationalists insist that two separate mechanisms are needed to learn regular forms and exceptions, while connectionists assert that a single mechanism can do both of these tasks.

Seidenberg acknowledges that the Rumelhart and McClelland past-tense model does indeed have many failings. He points out that this is the very first model of its kind, and so we should not be surprised to find that it is flawed. Connectionists are, however, eager to learn from their mistakes so as to improve their models. Seidenberg himself used the analysis of these mistakes to produce a new connectionist model of past-tense learning that avoids these failings.

Thus we see that throwing out the whole connectionist paradigm because of the failure of one particular early model would be like claiming that the Wright brothers' feat in flying the first heavier-than-air machine was worthless because it could carry only one person and could fly only for a short time. It is wrong to reject a whole new way of thinking because of its early shortcomings. As long as the proponents of the new ideas are willing to learn from their mistakes, and consistently produce new models that are superior to the old ones, there is every reason to believe that they will eventually come up with quite successful models showing that this approach is a fruitful one.

Conclusion

In sum, most of the criticism against connectionism comes from people who have worked for many years to elaborate alternative views of how the mind works. This is a common phenomenon in science – people who have spent their lives on an old theory, no matter how many faults it may be shown to have, find it difficult to see the value of a totally new approach. I have not detailed the criticism against the representationalist and the artificial intelligence views in this book, as that would require at least two more books. Moreover, books have been written on both of these topics. A variety of criticisms of the representationalist view of human cognition are presented in Benny Shanon's book, *The Representational and the Presentational*. A selection of reasons why the artificial intelligence approach is inadequate for understanding how the human mind works can be found in a book by Hubert and Stuart Dreyfus, *What Computers Can't Do*.

Thus there is no need for me to present these criticisms here. Instead, I will stress once again the most important differences between these approaches and connectionism. First, the earlier approaches do not consider the functioning of the human brain relevant for understanding the functioning of the human mind, whereas connectionism does. Second, connectionist theory provides an explanation not only for the successes of the human mind but also for many of its failures and errors. Third, this theory is much more complex than the other approaches because the human mind is, as I have mentioned, the most complex system in the known universe, so any attempt to explain it by some simple mechanisms can only be doomed to failure.

I hope you have enjoyed our journey through some of the intricacies of the human brain and the human mind. As this field is still new and in the process of development, some of what I have written here may be superseded as new theories and models are developed within the connectionist paradigm. Yet the basic understanding of this new way of thinking that I have tried to provide here should be useful in the attempt to follow new developments in the field as well. I have tried to present an approach, a way of thinking, rather than a list of facts, so as to provide an underpinning for learning more about this paradigm. The annotated references that follow are offered in this spirit, to give you an opportunity to deepen and broaden your understanding of this brave new theory.

Annotated references and suggested readings

Chapter 1

The subject of parallel distributed processing was developed most fully by David Rumelhart, James McClelland and their collaborators in the PDP Research Group. The classic work in the field is comprised of two volumes of their papers, entitled *Parallel Distributed Processing: Explorations in the microstructure of cognition* (Cambridge, MA: MIT Press, 1986). The first chapter of this work, "The appeal of parallel distributed processing," provides an excellent introduction to the entire subject.

If you are interested in studying connectionism in depth, a good introductory textbook is William Bechtel and Adele Abrahamsen's *Connectionism and the Mind: An introduction to parallel processing in networks* (Cambridge, MA: Basil Blackwell, 1991). This book takes a formal mathematical approach to the subject, in addition to discussing its philosophical implications, and thus requires a good background in mathematics. I have not found any introduction to the topic which does not require such a background; indeed, this is one of the reasons I decided to write the present book.

An excellent critique of the idea that minds are like computers may be found in Hubert and Stuart Dreyfus's book, *What Computers Can't Do: The limits of Artificial Intelligence*, 2nd edition (New York: Harper & Row, 1979).

Many books have been written about consciousness lately. If you prefer a scientific rather than a mystical approach to the topic, Daniel Dennett's *Consciousness Explained* (Boston: Little, Brown, 1991) is your best bet. Dennett's book is highly readable and often amusing.

Chapter 2

An excellent discussion of the various sorts of monism and dualism can be found in Mario Bunge's book, *The Mind–Body Problem: A psycho-biological approach* (Oxford: Pergamon, 1980).

An interesting speculation about the way in which our brains come to be structured differently in their fine details has been developed by Gerald Edelman. He presents these ideas rather technically in his book, *Neural Darwinism: The theory of neuronal group selection* (New York: Basic Books, 1987). A more popular formulation of these ideas, together with other, related notions, may be found in his *Bright Air, Brilliant Fire: On the matter of the mind* (New York: Basic Books, 1992).

Raymond Smullyan offers a highly amusing thought experiment about the possibility of constructing a machine that reads minds in "An episte-mological nightmare," chapter 6 of his collection, *5000 B.C. and Other Philosophical Fantasies* (New York: St. Martin's, 1983).

Jerry Fodor's argument for the impossibility of reducing the mental to the physical is found in the Introduction to his book, *The Language of Thought* (New York: Crowell, 1975). Most of the rest of this book is rather diametri-cally opposed to connectionism, as Fodor is one of the major exponents of representationalism (which is described briefly in Chapter 12 of the present book).

Chapter 3

Further details on the functioning of neurons and the transmission of neuronal messages can be found in any textbook on neurophysiology or biological psychology.

Chapter 4

One of the earliest semantic network models was presented by Collins and Quillian in their paper, "Retrieval time from semantic memory" (*Journal of Verbal Learning and Verbal Behavior*, 1969, Vol. 8, pp. 240–248), which details the experiment mentioned here.

Dennett's argument for a sort of connectionism is found in a technical philosophical paper called "Beyond Belief," in the collection *Thought and Object*, edited by Andrew Woodfield (Oxford: Clarendon, 1982, pp. 1–95). Reading the article as a whole requires familiarity with the philosophical literature on beliefs or propositional attitudes, but you can find the argument for connectionism on pp. 30–31.

Chapter 5

This chapter is based mainly on ideas presented in Rumelhart and McClelland's book (see the references for Chapter 1). Another important source of these ideas is the volume edited by Geoffrey Hinton and James Anderson, *Parallel Models of Associative Memory* (Hillsdale, NJ: Erlbaum, 1981), which was the pioneering academic work on the topic.

Edward deBono put forward his highly original outlook in his book *The Mechanism of Mind* (Jonathan Cape, 1969; Pelican, 1971), fully twelve years before the publication of Hinton and Anderson's collection. DeBono's book is a very readable presentation which contains several other analogies intended to help explain the workings of the mind, in addition to the light-bulb model.

Chapter 6

This chapter, like the previous one, is based mainly on the ideas presented in Rumelhart and McClelland's work (see references for Chapter 1).

Chapter 7

Much of this chapter is based on Geoffrey Hinton's paper, "Implementing semantic networks in parallel hardware," in Hinton and Anderson's collection (see references for Chapter 5).

Chapter 8

The description of a computer simulation of the way humans analyze two-dimensional scenes is based on a paper by P.H. Winston, "Learning structural description from examples" (AI Laboratory, Technical Report

231, MIT, Cambridge, MA, 1970). This simulation is an artificial-intelligence effort predating connectionist theory.

A detailed description of the NETTALK simulation can be found in a paper by Terry Sejnowski and Charles Rosenberg called "Parallel networks that learn to pronounce English text," published in the journal *Complex Systems*, Vol. 1, pp. 145–168.

The computer simulation of the way children learn the past tenses of verbs is described in chapter 18, "On learning the past tenses of English verbs," of Volume 2 of Rumelhart and McClelland's collection (see references for Chapter 1).

The model of the word-superiority effect and the experiments conducted on the computer program and on human participants to test the model are described in great detail in a set of two papers by McClelland and Rumelhart under the title "An interactive activation model of context effects in letter perception." Part I, "An account of basic findings," was published in *Psychological Review*, Vol. 88, pp. 375–407, in 1981. Part II, "The contextual enhancement effect and some tests and extensions of the model," was published in Vol. 89 of the same journal, pp. 60–94, in 1982.

The work on the arrays of odor receptors is reported in an article in *Science News*, April 10, 1999. Linda Buck's group is based at the Harvard Medical School, while Randall Reed's group works at the Johns Hopkins Medical Institutions.

Chapter 9

The distinction between what I am calling permanent and temporary memory is similar, but not identical, to one made by Endel Tulving, one of the most important memory researchers of our time, between what he calls "semantic memory" and "episodic memory." Tulving's definition of the distinction between these two types of memory, and his description of a long series of experiments designed to explore the second type, may be found in his book, *Elements of Episodic Memory* (London: Oxford University Press, 1983).

The new model of the distinction between temporary and permanent memory is described by James McClelland, Bruce McNaughton and Randall O'Reilly in their paper, "Why there are complementary learning systems in the hippocampus and neocortex" (*Psychological Review*, 1995, Vol. 102, pp. 419–457). This paper contains a survey of some of the most important ideas in connectionist theory, including many of those discussed in the present book. It is fairly readable and would provide a good place to start for readers interested in getting into the academic discussions of connectionism.

Chapter 10

You can read about the mental consequences of damage to the brain in any textbook on neuropsychology. The connectionist explanations of the ensuing problems and of the methods for circumventing them are mostly my own speculations.

Chapter 11

The ideas of Edward deBono's that are described in this chapter are discussed more fully in his early books, *The Mechanism of Mind* (see the references for Chapter 5) and *Lateral Thinking* (Ward Lock Education, 1970; Pelican, 1977), as well as in many other books he has written since. These books are very readable and interesting. They provide a wealth of suggestions for providing oneself with circumstances that may make perspective change possible, as this is best done indirectly.

Chapter 12

Jerry Fodor and Zenon Pylyshyn present their criticism in a paper entitled "Connectionism and cognitive architecture: A critical analysis," published in the journal *Cognition*, Vol. 28, pp. 3–71, in 1988. This paper is rather technical and requires a fair knowledge of cognitive science.

Dewdney's book, *Yes, We Have No Neutrons* (Wiley, 1997), is written on a very popular level. Aside from the chapter on connectionism, it provides several excellent examples of "bad science." If you are careful to follow Dewdney's guidelines and avoid his prejudices, you can learn a lot about how to avoid being taken in by claims that pretend to be scientific when they actually are not.

Pinker and Prince's paper, "On language and connectionism: Analysis of a parallel distributed processing model of language acquisition," appears in the same volume of the journal *Cognition* as the Fodor and Pylyshyn paper, on pp. 73–193. It too is a highly technical article, but it provides a good description of the contrasts between the representationalist and connectionist points of view.

Mark Seidenberg's discussion can be found in an article called "Connectionism without tears," which appears in the volume *Connectionism: Theory and practice*, edited by Steven Davis (Oxford University Press, 1992), on pp. 84–122. It too is fairly technical, but provides a good picture of how connectionists actually do their work.

Benny Shanon presents a wide variety of criticisms of representationalism from various points of view, stressing the narrowness and rigidity of this theory in contrast to the almost limitless flexibility of the human mind. His book, *The Representational and the Presentational* (Harvester Wheatsheaf, 1993), gathers together a great deal of empirical evidence discovered by many different researchers in the field. It is not too technical and should serve as a good introduction to the issues involved.

Dreyfus and Dreyfus's book is listed in the references for Chapter 1. Their criticism is probably not relevant to computer models of connectionist views, which are not discussed in their book.

Index